Fluid Rights

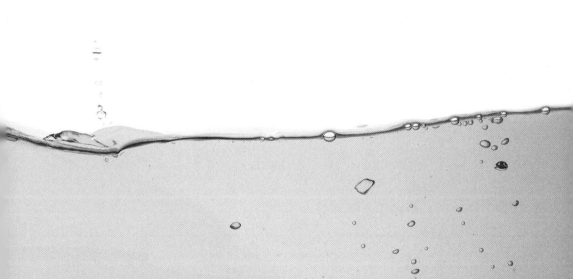

Fluid Rights

Water Allocation Reform in South Africa

Synne Movik

Published by HSRC Press
Private Bag X9182, Cape Town, 8000, South Africa
www.hsrcpress.ac.za

First published 2012

ISBN (soft cover) 978-0-7969-2353-0
ISBN (pdf) 978-0-7969-2354-7
ISBN (e-pub) 978-0-7969-2355-4

© 2012 Human Sciences Research Council

The views expressed in this publication are those of the author. They do not necessarily reflect the views or policies of the Human Sciences Research Council ('the Council') or indicate that the Council endorses the views of the author. In quoting from this publication, readers are advised to attribute the source of the information to the individual author concerned and not to the Council.

Copyedited by Glenda Younge
Typeset by Nicole de Swardt
Cover design by Firebrand

Distributed in Africa by Blue Weaver
Tel: +27 (0) 21 701 4477; Fax: +27 (0) 21 701 7302
www.oneworldbooks.com

Distributed in Europe and the United Kingdom by Eurospan Distribution Services (EDS)
Tel: +44 (0) 17 6760 4972; Fax: +44 (0) 17 6760 1640
www.eurospanbookstore.com

Distributed in North America by River North Editions, from IPG
Call toll-free: (800) 888 4741; Fax: +1 (312) 337 5985
www.ipgbook.com

Authored by Synne Movik (post doc)
Norwegian University of Life Sciences
P. O. Box 5003
1432 Aas
Norway

Contents

List of figures and tables viii
Acknowledgements x
Acronyms and abbreviations xii
Preface xiv

1 **Introduction** 1
  Water, scarcity and governance 1
  Allocation discourses 3
  South Africa: Pioneering water allocation reform 7
  The structure of the book 9

2 **Water rights in context: Evolution and reform** 11
  A tale of conquest 11
    The pre-colonial era 12
    The arrival of the Dutch and English 12
    Indirect rule 13
    Force and featherbed: The growth of mining and agriculture 15
    Deepening dispossession: The National Party's 'segregated development' 16
    A legacy of inequality 17
  The evolution of water rights regimes 18
    Roman-Dutch law comes to the Cape 18
    The British doctrine of riparian rights 19
    The capricious creation of colonial judicial policy 20
    The Water Act (No. 54 of 1956) 21
  Transition 22
    A fledgling democracy: Crafting policies for change 22
    The nature of the negotiated settlement: Protecting property 24
    The resurgence of traditional authority 25
    A brief review of the political economy context 26
  Getting the Act together: Processes and drivers 30
    The geophysical backdrop 30
    The initiation of reform 30

      Focus on services 33
      Environmental concerns 34
  The National Water Act 1998: Key features and debates 36
      Recognised water uses, the Reserve and Resource-Directed Measures 37
      Licences and trading 42
      Compulsory licensing 43
      Contested compensation: The 'safeguard clause' 44

3 WAR in the making: Crafting the Water Allocation Reform Programme 48
  Water Allocation Reform: The basis and the process 48
      Through the lens of scarcity 49
      The pillars of reform: Registration and compulsory licensing 50
  Processes, actors and perceptions 53
      The Department of Water Affairs and Forestry and the Department
          for International Development 53
      The Expert Panel and other actors 54
      Consultation and participation 56
  Emerging perspectives 58
      The industrialist/institutionalist perspective 58
      The agriculturalist/livelihoods perspective 59
  Water Allocation Reform: On paper 62
      From WARP to WAR 62
      Policy narratives and the construction of social identities 63
      Existing lawful users and historically disadvantaged individuals 64
      Privileged accounts 68
      Efficiency and equity 70
  Links to land: The absence of attention to acquisition 73
  Narrowing down the 'room for manoeuvre' 74

4 Water allocation in the Inkomati 78
  Context: Into the Inkomati 78
      Historical legacies shaping patterns of water use 79
      Sharing waters: The 1992 Agreements and the forging of new boundaries 84
  Characteristics of the Inkomati Water Management Area 86
      Geography and water availability 86
      Water management structures 91
  Patronage and paternalism 93
      Sugar in the Inkomati: A recipe for success? 93
      Emerging farmers 93

      Established farmers   106
      Perceptions of sharing and co-operation   109
  Contested allocations   112
      Scarcity and the blame game: Sugar versus non-sugar   112
      Inter-departmental struggle for allocative authority   116
      The little man against the State   121
  Preparing for compulsory licensing   122
      Validation and verification of water use   122
      Uncertainty and dynamics   124
      Outcomes   127
      Impasse   130

5   **Conclusions**   134
  Drawing it all together   134
  Emerging insights   139
      How discourses shape water rights   140
      The separation of land and water: Parallel processes, detached dynamics   144
      The difficulty of determining use   146
  What of the future?   146
      The status of WAR, and the emergence of new strategies   146
      Capacity constraints   147
      Challenges of regulation   148
      The politics of redistribution   150
      Wider implications   152

Appendices   155
References   165
Index   177

Figures and tables

FIGURE 4.1   Map of Nkomazi area   82
FIGURE 4.2   Inkomati Water Management Area (WMA)   88
FIGURE 4.3   CMA members in discussion at the Maguga Dam   95
FIGURE 4.4   Ronnie Morris and Mr Brown, a commercial farmer   95
FIGURE 4.5   Working in the sugarcane fields, Mzinti, Nkomati   96
FIGURE 4.6   Sugarcane fields stretching to the horizon, Madadeni, Nkomazi   96
FIGURE 4.7   Mr Nkalanga in his field at Spoons 7B, Komati River   96
FIGURE 4.8   The Khutselani Women's Group growing vegetables, near Driekoppies   96
FIGURE 4.9   Mr Edward Sambo in the pump house at Masibekela   97
FIGURE 4.10 Masibekela Farmers' Association next to the dam   97
FIGURE 4.11 Graphic representation of yield decline, Komati   98
FIGURE 4.12 Graphic representation of yield decline, Lomati   98

TABLE 4.1   1992 water sharing treaty   84
TABLE 4.2   Water yield and demand, NWRS 2004   88
TABLE 4.3   Water yield and demand, ISP 2004   89
TABLE 4.4   Water allocations to the Nkomazi and Mswati, in million m^3 116

In memory of my father

Acknowledgements

I WOULD LIKE TO express my gratitude to the Norwegian Research Council as well as the Department of International Environment and Development Studies at the Norwegian University of Life Sciences for their generous financial support. I wish to thank my supervisor at the Institute of Development Studies at the University of Sussex (UK), Lyla Mehta, for her generosity, continuous encouragement, and unfailing support throughout the process of writing this book. I also want to thank my local mentor in South Africa, Barbara van Koppen of the International Water Management Institute (SA Office), for generously sharing her knowledge and time, and for introducing me to a range of interesting and committed people in South Africa, and to IWMI staff for taking the time to discuss a variety of issues with me. I am indebted to my affable and very capable research assistant in the field, Ronnie Morris, and also to his lecturer at the Lowveld College of Agriculture, Dr Peter Reid, for enthusiastically sharing his experiences. A number of people took time off from their work, and generously shared their knowledge with me. I am grateful to many of the staff at the Department of Water Affairs and Forestry (South Africa), and the Department for International Development (UK) for sharing their knowledge and opinions with me. Thanks are also due to academics at the University of Pretoria, the Council for Industrial and Scientific Research, the University of the Witwatersrand and the Institute for Poverty, Land and Agrarian Studies (PLAAS) at the University of the Western Cape as well as people from Pegasys and the Komati Basin Water Authority. From all of these institutions, and from the regional office and CMA in Inkomati, I met interesting, professional and generous people. Particular thanks are owed to Gerhard Backeberg, Werner Comrie, Enoch Dhlamini, Geert Grobler, Brian Jackson, Francois Junod, Eiman Karar, Hadley Kavin, Kgomosoane Mathipa, Francois van der Merwe, Ndileka Mohapi, Bill Rowlston, Ash Seetal, Anthony Turton, Gavin Quibell, Maritza Uys and Dirk Versfeld. To the people of the Inkomati, to the emerging farmers, the dryland farmers, the commercial farmers and the extension officers – in particular Philemon Mthembu and Banie Swart – I am grateful. I much appreciated the welcoming attitude of the regional office people and the Head of Technical Support Services at the Department of Land and Agricultural Affairs. Many of the board members of the Catchment Management Agency kindly agreed to meet and spend hours discussing with me, as did people from AWARD and GeaSphere. I was extremely fortunate to be invited by MaTshepo Khumbane and Marna de Lange on a field trip to the Water for Food Movement's rainwater harvesting project. I learnt a lot from these two women, in particular the powerful MaTshepo, for whom I nurture a deep respect. I am grateful to Eva Masha, for housing us in her own home. Thanks also to colleagues and fellow students at IDS, especially Kattie, for her many kindnesses. I wish to thank Ian Scoones and Desmond McNeill for their constructive feedback, and Sunetro Goshal at Noragric who generously offered to read the draft manuscript for this book – I am very grateful for his incisive and constructive comments. I would also like to thank the two anonymous reviewers who provided very useful feedback. I am very lucky to have had the support, flexibility and understanding of my husband, Ole Petter, which made this research possible. And to my little radiant, warm-hearted daughter Mathilde, who handled being shifted around from Norway to the UK to South Africa with ease and good humour.

Acronyms and abbreviations

AgriSA – Agri South Africa
AMD – acid mine drainage
ANC – African National Congress
AWARD – African Water Issues Research Unit
AWIRU – Association for Water and Rural Development
BBBEE – broad-based black economic empowerment
BEE – black economic empowerment
CASP – Comprehensive Agricultural Support Programme
CGIAR – Consultative Group on International Agricultural Research
CSIR – Council for Scientific and Industrial Research
DALA – Department of Agriculture and Land Affairs (Mpumalanga)
DBSA – Development Bank of South Africa
DFID – Department for International Development (United Kingdom)
DNA – Department of Native Affairs
DLA – Department of Land Affairs
DoA – Department of Agriculture
DWAF – Department of Water Affairs and Forestry
ELU – existing lawful user
FDI – foreign direct investment
FWARCS – fractional water allocation and reservoir capacity sharing
GA – general authorisation
GDP – gross domestic product
GEAR – growth, employment and redistribution
GIS – geographic information system
GNI – gross national income
GWCA – government water control area
HDI – historically disadvantaged individual
IPILRA – Interim Protection of Informal Land Rights Act
ISP – internal strategic perspective
IWMI – International Water Management Institute
IWRM – integrated water resource management
JWM – Joint Water Commission
KOBWA – Komati Basin Water Authority
KRIB – Komati River Irrigation Board
LRAD – land redistribution for agricultural development

MAFF – Ministry for Agriculture, Forestry and Fisheries
MAFU – Mpumalanga African Farmers' Union
MCCAW – Mpumalanga Coordinating Committee for Agricultural Water
MCGA – Mpumalanga Cane Growers' Association
MEC – Member of the Executive Council
MRDLR – Ministry for Rural Development and Land Reform
NAFU – National African Farmers' Union
NIEP – Nkomazi Irrigation Expansion Programme
NGO – non-governmental organisation
NOWAC – Nkomazi/Onderberg Water Action Committee
NWA – National Water Act
NWRS – National Water Resources Strategy
PIU – Project Implementation Unit
PLAAS – Programme for Land and Agrarian Studies
PTO – permission to occupy (also known as 'RTO' - right to occupy)
RDP – Reconstruction and Development Programme
SACP – South African Communist Party
SADC – South African Development Community
SADT – South African Development Trust
SASA – South African Sugar Association
SASEX – South African Sugar Extension Service
SASRI – South African Sugar Research Institute
SLAG – settlement/land acquisition grant
TLGFA – Traditional Leadership and Governance Framework Act
TSB – Transvaal Suiker Beperk (Transvaal Sugar Company), now TSB Sugar
UNDP – United Nations Development Programme
WAMI – Water Allocation Monitoring Index
WAMS – water administration and measurement system
WAR – water allocation reform
WARMS – water authorisation and registration management system
WARP – Water Allocation Reform Programme
WfGD – water for growth and development
WFM – Water for Food Movement
WFSP – Water and Forestry Support Programme
WMA – water management area
WRC – Water Research Commission
WRM – water resource management
WRYM – water resources yield model
WUA – Water User Association
WULATS – water use licence application tracking system

Preface

HOW ARE RIGHTS TO water resources allocated? How do practices, doctrines and policies evolve that shape the way in which water use rights are conceptualised, governed and accessed by women and men, the State and its citizens? When the authority to allocate water is vested with the State, what stories are told at the policy level to justify the choice of particular mechanisms and patterns of allocation over others, and how do these understandings mesh with local-level discourses and practices? Such questions are at the centre of this book.

The perceived threat of water scarcity and the rise of the Integrated Water Resources Management paradigm have prompted policy reforms worldwide. South Africa's Water Act of 1998 is regarded as being particularly progressive, not least due to its explicit focus on redressing past injustices through the redistribution of water resources. The Act did away with the system of riparian rights and instead introduced administrative water rights which outline the main principles of allocation. These principles needed to be fleshed out in a more practice-oriented policy, however, which gave rise to the Water Allocation Reform strategy. Through these processes, the State was vested with a large degree of discretion to determine use rights and criteria for allocation, which has created space for the emergence of what I term *allocation discourses*.

I have adopted a two-pronged strategy in order to examine these allocation discourses: I carefully analysed the reform processes and discourses and then compared and contrasted these with a case study of how these processes were being played out in the Inkomati Water Management Area. Analysing policy reform as discourse allows one to tease out particular assumptions of equity and efficiency, and see how causal relationships, which serve to shape the way water use rights are defined and allocated, are represented and justified. In this book I argue that in the early days the reform process essentially entrenched the powers of existing users and narrowed down the State's room to manoeuvre. Intra- and interdepartmental struggles for authority, a lack of capacity, and the difficulties of determining the extent of existing uses were other factors which led to reform efforts ending in a temporary impasse. A central feature of the argument is the notion that use rights and allocation patterns are created through discursive struggles at multiple levels.

The material for this book was gathered during fieldwork in 2006, with brief follow-up visits in 2009 and 2010. I draw on interviews with policy-makers, lawyers, water practitioners and academics, as well as a range of actual and potential water users. I also interviewed staff at the Catchment Management Agency, staff at the Department of Water Affairs headquarters in Pretoria and at the regional office in the Inkomati, the Provincial Department of Agriculture and Land Affairs, as well as other stakeholders. The problem with books about particular policies is that the world keeps moving, and today's fresh insights may become tomorrow's dated knowledge. However, I believe that the insights that emerge from this book are widely applicable, not only in the South African context, but wherever water reform is an issue.

I

Introduction

WATER IS A VITAL resource, which is fundamental to human life and wellbeing, ecosystem health, agriculture, energy and industries. It is at the heart of human existence, society and development. Attention to water resources is mounting the world over, and frequent statements contending that we are in the midst of a water crisis forcefully drive home the point that water is becoming increasingly scarce and should not be squandered. Emerging global visions of water, rife with universalised sound-bites, such as the 'gloomy arithmetic of water' (World Water Commission in Conca 2006: 151), are underpinned by dry statistics like the Water Poverty Index (Lawrence, Meigh et al. 2002). These gloom-filled narratives of dwindling resources have only recently begun to be challenged, with scholars and activists pointing to the multiple and often neglected dimensions of scarcity, and how it is deeply connected with issues of power and access (Mehta 2005; UNDP 2006; Mehta 2010). Nevertheless, the scare of scarcity has brought much attention to the issue of water governance.

Water, scarcity and governance

Water governance is essentially about people: how they relate to one another as individuals, groups and nation-states across different spatial and temporal scales, and how these relations are shaped by geographical and physical characteristics, the nature of the water sources, and constellations of power and authority. Water governance is multidimensional, encompassing government, civil society and the private sector, and involves a range of social, economic and administrative systems to develop and manage water resources and the delivery of water services. It is characterised by a diversity of institutions and stakeholders, often with competing and contested claims, offering the potential for both conflict and co-ordination in the access and use of water.

One important element of water governance is the idea of *property rights* or *use rights*, in order to regulate access to and use of water. Property rights are basically social relations, that is, they are 'humanly devised constraints that shape interaction' (North 1990: 1). However, access to and use of water is complex and messy, as water is a changeable and ephemeral element, and essentially the exclusive possession of water is impossible (Getzler 2004). When water is plentiful and there is enough to go around, water is regarded as a natural entitlement and a public good which all can access and use according to their needs. When there is scarcity of water and when demand increases, however, competition becomes stronger, and particular arrangements or doctrines emerge that serve to control access. One such arrangement is the *doctrine of riparian rights*, which is based on the principle that those who own land adjacent to a flowing stream/river (riparian land) are entitled to use water *reasonably*, in other words, in a way that does not infringe on the ability of any of the other riparian landholders to enjoy their right to reasonable use. This doctrine implies that only a certain group of people – namely those owning land bordering on the river – should be granted access to the water in that river. Another arrangement is the *prior appropriation doctrine*, which holds that the first to access and abstract water from a given source gains the right to that water, so it is basically a 'first in time, first in right' principle. Such doctrines evolve over long periods of time in different contexts, and help bring some stability and assurance to people's actions and patterns of water access. These different stages or doctrines do not necessarily follow in succession, however, and they may also co-exist and overlap, as is the case, for example, in parts of the eastern United States (Rose 1994b). Other, less formal, systems of water use have emerged in many developing country contexts, which are referred to rather loosely as 'customary' systems of water rights, where communities over time evolve particular rules and ways of engaging in water management in accordance with their particular customs, practices or religious beliefs (Van Koppen, Giordano et al. 2007). Such formal and informal systems may co-exist in the same socio-political space, giving rise to the notion of 'legal pluralism' (Von Benda-Beckmann 2001; Roth, Boelens et al. 2005; Van Koppen, Juma et al. 2005).

Over the last couple of decades, there has been a strong trend towards bringing water resources under the ambit of the State (Burchi 2005). The drive towards having the State control a country's water resources can, in part, be attributed to the hugely influential paradigm of Integrated Water Resources Management (IWRM), which emphasises the holistic management of water and the desire to do away with a piecemeal, sector-based approach to water management. Perceptions of scarcity and the rise of the IWRM approach have been the main drivers of legislative and policy reform across the globe, and since the late 1990s, governments in both industrialised and developing countries have embarked on comprehensive reform processes of their water sectors to overhaul the institutional and legal frameworks for water governance (Saleth & Dinar 2000; Garduño Velasco 2001; Saleth & Dinar 2004; Burchi 2005; Conca 2006). Examples of countries that have reformed their legislations include

Chile (1980), Mexico (1992), Uganda (1995), South Africa (1998), Sri Lanka (2000), Ghana (2001), Namibia (2004) and Australia (2007).

A key feature of many of these reform efforts has been the *institutionalisation of formal water rights* as a way of facilitating allocation of water between users, as has been the case, for example, in Australia, Chile, Mexico and South Africa (Bauer 1997; Garduño Velasco 2001; Godden 2005). Such tradable water rights have taken a variety of forms, such as grants, concessions, permits or licences, but basically they share the same key attribute, namely that they are initially issued by the State. Formal water-use rights administered by the State represent a break with previous legal doctrines such as riparian rights and prior appropriation. Some view a system of transferable, or tradable, use rights as a straightforward way to deal with scarcity. The key argument is that allowing 'the invisible hand' to work unobtrusively will best promote efficiency, although it acknowledges that formal rights represent an 'unhealthy commodification of a public good' (World Bank 2003: 16). Tradable use rights are nevertheless endorsed, as the idea is that the 'unhealthy' dimension will be offset by the increased tenure security, efficiency benefits of transferability, and associated reduced transaction costs. Critics of the tradable licences approach (see, for example, Roth, Boelens et al. 2005; Van Koppen 2006) hold that this view does not pay adequate attention to the existence of informal water rights and how people engage in water management in informal or customary ways that are not necessarily acknowledged by official laws. Others stress the notion of public trust and the authority of the State in issuing use rights, and argue that this approach provides an opportunity to facilitate redistribution and thereby ensure greater equity among users. The State is portrayed as the custodian of the commons (Stein 2000; Woodhouse & Chhotray 2005; Stein 2006), and as a custodian, it is supposed to act in the public interest and ensure equitable, efficient and sustainable allocation of water. This is particularly true of South Africa. The country's turbulent past, characterised by colonialism and apartheid, has caused huge levels of inequality in access to water resources. The State is seeking to redress these inequalities in access through its water reform policy efforts, and the country's water legislation has been lauded as one of the world's most progressive because of its emphasis on equity. South Africa thus presents an interesting case in terms of examining notions of equity and efficiency in the allocation of water rights.

Allocation discourses

Bringing water under the authority of the State, and the institutionalisation of formal use rights, opens up spaces for negotiation over the meanings and attributes of water-use rights in the formulation of legislation and policy. How are principles and guidelines for allocation formulated and interpreted? How are reasons arrived at for allocating water to this use rather than that use? Policy is not value neutral. Embedded in policy texts there will be certain understandings and perceptions that depend on particular assumptions rather than others. Treating policy as a set of discourses, and analysing these in order to understand and tease out hidden assumptions and values, is a means of contending with the failings of empiricist policy analysis, which has long tried to emulate the procedures of the hard sciences by operating with notions of objectivity, making sharp distinctions between facts and values, and maintaining value neutrality. Rather than trying to grapple with reality itself, analysing policy as discourse deals with the *accounts* of reality that are to be found in policy. It is not the objects or the properties *per se*, but rather the 'vocabularies and concepts used to know and represent them that are socially constructed by human beings' (Fischer 2003: 217). These vocabularies and concepts represent a form of power in that they constrain, but also enable, particular ways of acting and being (Foucault 1980; Foucault 1991). Through the process of framing, policy discourses create certain categories and labels that open up some avenues for action, whilst closing down others. Framing is what happens when policy makers highlight some features of a perceived problem and downplay others according to their particular 'ways-of-seeing' (Hajer 1995; Fischer 2003) and involves '…selecting, organizing, interpreting and making sense of a complex reality to provide guideposts for knowing, analysing, persuading and acting' (Rein & Schön 1993: 146). The process of making sense involves the creation of certain storylines or narratives that provide actors with a 'symbolic set of references that suggest a common understanding' (Hajer 1995: 62), and such storylines or narratives are characterised by construing a logical chain of causation, by having a beginning, middle and end (Roe 1991; Kaplan 1993; Roe 1994). The framing of a policy issue, and the emergence of particular policy narratives, is always part of some broader political and economic setting, which in turn is located in a particular place and historical period.

Analysing the ways in which relationships of cause and effect are framed has been used to great effect to explain particular environmental problems. Stating that environmental problems are socially constructed has almost become a 'platitude', according to Hajer (1995: 42). Hajer's seminal work on acid-rain demonstrated how it was not the objective, measurable facts that were the problem, but rather the ways of seeing the facts that became important. Less attention has been devoted to how policy applies *use rights* to environmental resources. This is the case in spite of the fact that environmental problems have been commonly regarded as resulting mostly from inappropriate legal systems and tenure arrangements (Repetto & Gillis 1988; Jänicke 1988; Weale 1992 cited in Berkhout, Leach et al. 2003). Property rights, like all social phenomena, 'abstract from the real world by *stylizing select characteristics* of human behaviour, organization, and physical environments' (Eggertsson 1996: 157, emphasis added). It is the nature of this stylising that is of interest, and how it gives rise to certain narratives or storylines. How are potential users – and, consequently, use rights – stylised in national water policy? The forging of what can be called 'subject positionings', that is, particular ways of talking about users in policy narratives, define certain categories and their relations vis-à-vis one another and the resource (Fairclough 1995; Laclau & Mouffe 2001; Fischer 2003). I argue that language in this context is not peripheral; rather, contestations over the meanings of use rights are played out through language.

In undertaking discourse analysis, one of Foucault's critical aims is to beware of interpretation. I think that researchers and authors should try not to add their own layer of meaning on top of presentations of discourse. However, although authors should strive for this as an ultimate objective, it is almost impossible to achieve in practice, since the author as observer influences the material by the very essence of her (or his) class, ethnicity, race, gender, mind-set, upbringing and cultural context. Consciously or not, the researcher is influenced by all these factors as '[w]e cannot study lived experience directly, because language of speech and systems of discourse mediate and define the very experience we attempt to describe' (Denzin & Lincoln 2003: 51). In actively selecting, interpreting and assembling, we not only tease out what is there, but we are also intrinsically active in shaping a new product that we can hold up to the light and then exclaim, 'See what I found!' Reflexivity thus involves scrutinising the process itself, and asking questions such as 'What kinds of overt or covert criteria are used in selection?' and 'What influences are shaping the mind of the analyst?' As nothing is value free, so, too, is the beholder's mind full of anticipations and presumptions, and these are brought forcibly to bear on the identification of discourses. Researchers are thus always struggling with what Graaff terms the 'pernicious influence of discourse' (Graaff 2006: 1391). It is easy enough to imagine the resultant problems. For example, the chances of replicability, of someone being able to go into a similar project of discourse analysis and arrive at more or less the same conclusions, are highly unlikely. At best, researchers may find that major points in similar projects of discourse analysis do mirror one another. I bring with me my cultural baggage to my meetings with the policy makers, the Afrikaners, the rural people, extension workers, and so on. My ideas about them, built up through reading and meetings and thinking, slot them into a certain frame. Often, I found that I was battling with old presumptions and more or less covert prejudices in certain encounters with people, especially with some of the Afrikaners whom I met during fieldwork. I had to consciously make an effort to rid my mind of ossified ideas, try to be as open and *empty* as possible, and thus observe the perspectives evolving through interactions and interviews. The asset of the discourse-theoretic approach is that the researchers strive not to impose themselves into the act of interpretation, and attempt as far as possible to regard the elements of discourse at a distance. Nonetheless, my project of exploring the discourses around water-use rights will yield certain insights that are shaped by my own experiences.

Offering a discourse analysis of policy formulation may prove useful in understanding how certain concepts create certain storylines and narratives rather than others, and in teasing out how these become established. However, what about policy *in action*? What practical implications do particular policy formulations have? Of significance is not only how policy discourses are produced at the national level, but also how they are 'reproduced and transformed in a particular set of practices' (Hajer 1995: 44). Policies are created at specific historical points in time and in specific socio-economic settings that affect the ways in which they are conceived as discourse and practice. Context is not just the receiving container for policy, but according to Lave, the 'social world [is] constituted in relation with persons acting' (Lave 1993: 5, quoted in Wagenaar & Cook 2003: 147–148). These 'persons acting' also engage in discursive practices of their own, so context is not a blank page on which policy discourses are inscribed. What must be taken into consideration, then, are not only the discursive practices – how water users are talked about and defined – at the national level, but also the discursive practices that exist in the spaces in which policy unfolds, which is where it is implemented. This research process is not equivalent to policy appraisal or evaluation. Rather, it tries to see how discourses at national and regional levels compare and contrast, and to see to what extent they differ and in what respects. In other words, how do the storylines created at the national level resonate on the ground? How do these understandings mesh with local-level discourses and practices, and what are the implications for the reform process?

Whereas much research effort has been devoted to studying how rights to common-pool resources such as water are negotiated at the local level (see for example Bruns & Meinzen-Dick 2000), less attention has been paid to how ideas about rights are fashioned at the policy level, and how the interaction between policy and ground-level discourses shape conceptions of use rights and patterns of allocation. The global trend of water rights reform therefore warrants closer attention. It becomes important to understand better how power works through policy-level discourses and how these compare and contrast with local-level understandings. The originality of this approach lies in the fact that it perceives negotiations over the meanings and purposes of rights and allocation mechanisms as occurring not only at local levels between individuals and groups, but also *discursively* at the policy level.

In the shift to an administrative licence-based system of water allocation, the onus is now on the State to flesh out conceptions of use rights and to justify particular allocation mechanisms. This shift to State *allocation authority* creates room for the emergence of particular narratives relating to how water-use rights should be understood and allocated. The language of policy-making actively shapes the concept and content of use rights, and the description of actual and potential rights holders vis-à-vis these concepts and how these are expressed through policy and practice. Key questions that emerge, then, are: How do ideas and interpretations of water-use rights manifest themselves in policy formulation, and how are they contested? How do understandings of key notions such as scarcity, equity and efficiency shape conceptions of use rights and allocation mechanisms?

South Africa: Pioneering water allocation reform

In 1998, South Africa ratified one of the most sophisticated and progressive water acts in the world: the National Water Act (No. 36 of 1998). Here was a country, newly emerged from decades of oppression, which was trying to embark on a fresh start. To do this, a vital task that the new government needed to address was how to share the country's water resources among its people. It is a particularly interesting case to explore in that it explicitly emphasises equity as being the primary objective for instituting a water-use rights reform. Though efficiency is a key concern, the decision to reform water-use rights and institute licences through mechanisms such as 'compulsory licensing' has primarily been framed in terms of facilitating a redistribution of rights to redress a highly skewed distribution of water resources. The redistribution of water is thus also seen as key in terms of poverty eradication.

South Africa's legacy of inequality can be traced back to the early history of colonial conquest by the Dutch in the seventeenth century, followed by the British in the early eighteenth century. The exploitative nature of colonisation deepened and hardened in the modern era, through more than 40 years of segregationist and deeply repressive policies under the apartheid regime that began in 1948 with the coming to power of the Afrikaner nationalist, DF Malan. Apartheid's ideology of segregating peoples according to their skin hues left indelible marks on the South African national psyche. The ANC was formed in 1912 to promote the rights of black people. Banned by the apartheid government in 1960, it moved underground and became the centrepiece of African resistance to the repressive rule of the apartheid regime. This historical trajectory resulted in South Africa becoming a country of 'plenty amidst poverty' with staggering levels of inequality (Nattrass 1983: 12, quoted in May 2000: 16).

The ending of apartheid with the transition to democracy in 1994 saw the inauguration of Nelson Mandela as the country's first democratically-elected president. A new era had begun. The Constitution of the Republic of South Africa,

1996, enshrined equality and respect as cornerstones of the fledgling Rainbow Nation, and reforming the country's various legislations, particularly those governing resource access, became a key concern for the new government. As water was perceived as a scarce resource that was very unequally distributed, water legislation was high on the agenda and was one of the first to be overhauled and aligned with the principles of the Constitution.

The National Water Act (NWA) recognised water as a human right, and instituted the concept of a Reserve that covered both the human right to water as well as ecological needs. The guiding principle of the Act is the need to redress past inequities through the redistribution of formal water-use rights (compulsory licensing), which need to be registered and paid for. Another key concept is the decentralisation of management through the establishment of Catchment Management Agencies (CMAs). The NWA did not deal specifically with how water rights should be distributed, however, and remains unclear with respect to how rights should be interpreted (Perret 2002). It speaks of achieving optimal allocation of water resources, and encouraging the 'beneficial use of water for the public good', but it is by no means unanimously agreed upon what these concepts entail or how they should be implemented.

The fact that the authority to issue time-limited use rights was now squarely placed under the aegis of the National Department of Water Affairs and Forestry (with the introduction of a nationwide registration system) raised a key paradox. On the one hand there was a move towards decentralisation with the institution of the CMAs, whilst on the other hand, the 'nationalisation' of water use through a nationwide registration system was a move towards centralising authority in terms of issuing use rights. This paradox reflects the different views of formal rights as described earlier, and gave rise to different perspectives on how rights should be conceived. The objective of this book is to offer an understanding of the dynamics of South African water rights reform through an examination of the ways in which water-use rights and allocation mechanisms have been conceptualised and shaped through policy and practice.

The legacy of inequality left by apartheid laid the ground for the focus on redress and redistribution, but the nature of what became known as the negotiated settlement under the transition (where the ANC was thought to reach a compromise with powerful economic interests in order to secure political power) paved the way for the rise of the neoliberal paradigm in much of post-apartheid politics. This paradigm influenced water reform in various ways, such as in the heavy emphasis placed on the role of water for economic productivity and the recognition of existing lawful uses, which was justified by the fear of disrupting the economy. It is also reflected in the emphasis on the notion that those who had been wronged in the past should be given access to water, but with the general ideal that they should become commercial and economically productive users of water. A parallel discourse, deeply distrustful of the State's ability to achieve redress, focused on the role of rural communities in terms of their scope for self-management of water resources, particularly for agriculture.

Whilst these discourses struggled for space at the national level, the picture at the regional and local levels deviated from some of the simplistic assumptions they contained. The narratives and discourses at the regional and local levels did not conform to those promoted at the national policy level, and undermined some of the assumptions of policy actors. Again, an understanding of the historical trajectory is a key aspect for appreciating the particular positions and constellations that were present in the Inkomati region, in particular the role played by traditional authorities.

The structure of the book

What follows is a brief explanation of the structure of this book.

Chapter 2 describes the larger context within which South Africa's water policy reform is taking shape. This chapter highlights how history provides a key impetus for the current reform, and how past practices and the brutal policies of apartheid segregation combined with colonial doctrines to lay the foundations for the current skewed system of resource distribution. It offers a concise overview of the last 300 years of South African history, putting in historical context why water reform is so central to the country's development and future prospects. This chapter describes the evolution of water rights regimes with the shifts in legal basis for allocating water rights over time. The chapter then moves on to cover the more recent period and the build up to the reform of the water legislation, situating these efforts in the particular political economy context of policy reform in the 1990s. It highlights the socio-political nature of the transformation of the country from an isolated siege economy to a non-racial democracy and a global world player, and places the water reform efforts within the turbulence of macroeconomic policy shifts and the resurgence of traditional authority. The key features of the Act itself are described, highlighting contestations and debates around core issues.

Chapter 3 focuses on the making of the policy (the Water Allocation Reform) by analysing the political, personal and professional manoeuvrings that went on during the construction of the policy. It describes the process of drafting the guidelines for the Water Allocation Reform, which was intended to clarify, put flesh on and make practicable the principles outlined in the Act, and identifies the two main competing discourses reflecting wider divisions in policy and politics. The chapter concludes by drawing links between the water reform process and the experiences from the land reform.

Chapter 4 provides a local case study, which looks at water allocation in the Inkomati region, and focuses in particular on the way it complicates the analysis of the national-level policy process. The voices of real people on the ground come forward in this chapter, including departmental officials, white farmers and emergent farmers. In the committee rooms of Pretoria, one version emerges, but this is transformed, confused and obfuscated at the local level by new processes of politics and power, where new discursive practices (and new 'positionings') emerge. This argument reinforces the general argument about rights and discourse, but makes it more sophisticated. The chapter then goes into more detail on the actual reform processes emanating from the national policy narratives. The bestowing of entitlements could only be achieved through a process of validation and verification that would define the content of the use right. This process was made difficult because of scientific uncertainties in terms of determining how much water people were actually using, the messy nature of land access, and the strong sentiments of the white farmers as representing the 'little man against the State', and their associated reluctance to co-operate with reform efforts. Hence, it proved a complex task to determine the extent of existing lawful use rights scientifically and legally which, in turn, threatened to jeopardise the whole water allocation reform process.

The concluding chapter draws together the insights from the previous chapters. It offers reflections on the wider political economy of process and outcomes, linking up to other parallel processes such as land reform – both in terms of the influence this had on what happened, and the implications for wider political-economic change in South Africa. What does this tell us about the political economy of South Africa, and what tensions and conflicts might emerge in the future? The concluding chapter also reflects on lessons from South Africa for the wider experience of water policy and legislation, situating the book in a broader international context.

2

Water rights in context: Evolution and reform

THIS CHAPTER SETS OUT to provide the historical context of modern water allocation reform in South Africa. It describes how water rules and formalised laws took shape during the time of conquest and colonialism, and later on in the apartheid era. It emphasises how the discriminatory land laws, in particular the Natives Land Act[1] and the creation of the homelands, which served as reservoirs of cheap migrant labour for industry and the mines, and the heavily subsidised commercial agricultural sector shaped the terrain in which aspiring African peasants had little chance of surviving. The chapter then goes on to describe the evolution of water doctrines and legislation, highlighting how water came to be linked to land possession, and how a hybrid system of riparian rights and permits developed prior to the promulgation of the new National Water Act (NWA). The transition to democracy in 1994 ushered in a whole new era, and gave rise to hopes of redressing past discriminatory practices. It saw the State wielding unprecedented authority in terms of administering water-use rights, and the setting about of realising redistribution within a context of a rapidly shifting political landscape.

A tale of conquest

The evolution of water law must be placed in the wider historical context of South Africa's history, which is one of oppression, dispossession and the suffering of three and a half centuries of protracted struggle for dominion, characterised by increasingly brutal conflicts along ethnic and racial lines. I will outline in brief the key features of South Africa's history before going on to examine the evolution of water laws in more detail.

The pre-colonial era

South Africa is an extremely diverse country, in all senses of the word. It encompasses a multitude of peoples, from the nomadic hunter-gatherer San and pastoralist Khoikhoi peoples of the south-western Cape, to the Bantu-speaking peoples who arrived from the Niger Delta, and settled in what is today's South Africa. Of the Bantu tribes, the Nguni tribes (Zulu, Swazi, Xhosa and Ndebele) largely settled along the eastern seaboard, the Sotho-Tswana peoples (Tswana, Pedi and Basotho) stayed in the Highveld areas, and the Shangaan-Tsonga peoples, along with the Venda and Lemba, settled in the north-east.

Chiefly rule and kinship groups were characteristic features of traditional institutions in pre-colonial South Africa. Although tribes 'organized under the domination of elders, they contained redistributive mechanisms that thwarted tendencies to reproduce inequalities in a cumulative fashion' (Mamdani 1996: 41), the nature of the relationship between elders and their subjects is a matter of debate. Chiefly power was checked by clan- and lineage-based village councils, which regulated access to land, forage and other natural resources, and dealt with other matters such as trade and family obligations. The regulation of social and economic affairs by village-based communities was, therefore, the backbone of traditional society. But the nineteenth century turned out to be a turbulent period which was hardly conducive to the 'stable reproduction of customary relations' (Mamdani 1996: 42). Nguni conquests gave rise to the *Mfacane*[2] or *Difaqane*, in which the Zulus, under the leadership of the legendary militaristic innovator Shaka Zulu, conquered many of the neighbouring clans, who fled in terror. The *Difaqane* wrought havoc in the region, weakening resistance to the colonial conquest of the hinterland.

The arrival of the Dutch and English

The arrival of Jan van Riebeeck and the Dutch East India Company in the Cape in 1652 marked the beginning of a long, violent history of conquest and marginalisation, which has left indelible marks on modern South Africa. The Dutch were not to remain alone in the Cape[3] for long, however, as the British arrived on the scene as the eighteenth century drew to a close (Sparks 2003b). A protracted jostle for power ensued, with the British finally gaining sovereignty of the Cape in 1814, and in 1845 declaring the eastern seaboard around Natal a British colony (Terreblanche 2002). Frustrated with the British power grab and dismayed by the new ways introduced by the British in terms of ownership of land and labour, the Boers[4] resisted by moving inwards in search of new land. Thus began in the 1830s what became known as the Great Trek, which eventually resulted in the establishment of two independent Boer Republics in the hinterland north of the Orange River – the Orange Free State and the Transvaal (Terreblanche 2002; Sparks 2003b). The Boers aimed to recreate in their new states the 'labour patterns, property relations and patriarchal feudal order that prevailed before the arrival of the British', and in this they succeeded remarkably well (Terreblanche 2002: 219).

The Boers did not find it too hard to deprive the African tribes that resided in the inland areas of their land, arriving as they did in the midst of the *Difaqane* (Morris 1998; Terreblanche 2002; Sparks 2003b). The Boers ended up taking most of the land, meeting only small pockets of resistance from the Venda and the Pedi, who held out for some time. Moshoeshoe, the leader of a Sotho tribe, fled with his people up the mountainous hillsides into what was eventually to become Lesotho, and sought British protection from the Boers, as did the Swazis and Tswana (Sparks 2003b: 107).

The continuing battle for control between the Boers and British culminated in the Anglo-Boer War of 1899-1902, which was the 'greatest war the African continent had ever known' (Sparks 2003b: 124). The outcome of this war was that

> the Boers lost the war, but in a wave of reparative sentiment for the perceived injustice done to a small nation by a powerful one, they won the peace. And in that victory their exclusionist credo – the notion that the land was theirs by divine right and that the indigenous blacks were aliens in it, whose role was to serve but not participate – was extended from the two Boer republics over the whole of South Africa. It triumphed over the more inclusionist liberalism that had been established in the two British colonies, and became national policy in a new independent country created out of a union of all four territories.
> (Sparks 2003b: 124).

Thus, with the creation of the Union of South Africa in 1910, the British had the economic clout, the Boers had the political power, and the large African majority was increasingly alienated in its own land.

Indirect rule

Whilst the Boers were initially largely satisfied with their 'bitter almond hedge' strategy (Sparks 2003b), which meant physically separating themselves from the Africans, the British adopted a more active tactic through the institution of 'indirect rule' (Mamdani 1996; Morris 1998). Indirect rule was predicated on the belief that Europeans and Africans were culturally distinct, and the institutions of governance most suited to Africans were those that had been constructed within the webs and skeins of tradition (Mamdani 1996; Oomen 2005). Colonial rule proceeded by governing through these local institutions, rather than completely replacing them with colonial interventions. '[I]n practice, indirect rule laid heavy emphasis on the role of the chief in the government of African peoples, even for those peoples who traditionally did not have political as distinct religious leader' (Crowder 1968: 169, in King 2005: 65). Indirect rule gave a great deal of power to the chiefs, as the government structures were stripped of the traditional checks and balances that used to be in place to curb their supremacy. Increasingly, the chief often became an instrument of the colonial government, with the apartheid era officials seeing to it that uncooperative chiefs were removed. The British deposed and marginalised rebellious chiefs while rewarding those that did their bidding, thus exploiting competition, fostering uncertainty and sowing the seeds of discontent for future generations (King 2005). Ntsebeza (2000) suggests that the use of traditional authorities by colonial powers enabled them to exploit ambiguities in the relationship between chieftaincies and their people, particularly in terms of the accountability and legitimacy of traditional systems. For example, the British freely appointed leaders without consulting councillors or elders, sometimes choosing leaders that were not in direct line of lineage. In 1894, the British passed what became known as the Glen Grey Act, which strengthened the power of traditional authorities at the local level. This move was instrumental in shifting land ownership in nature areas (non-agricultural and non-industrial areas) from communal to individual tenure, while leaving the chief with authority over unallocated land (Hall 2004). After the conciliation of the Boers and the British was manifested through the creation of the Union of South Africa in 1910, the pace of discriminatory law-making quickened. In 1913, the notorious Land Act was passed, which effectively reserved a minor and marginal portion of the land for Africans. Sol Plaatje, a Tswana-speaker of the Baralong tribe and first in the 'long line of fine African writers' (Sparks 2003b: 417) described the effects of the Act in the following terms: 'South Africa has by law ceased to be the home of any of her native children whose skins are dyed with a pigment that does not conform to the regulation hue' (Sol Plaatje, quoted in Sparks 2003b: 135–36).

The Native Trust and Land Act (No. 18 of 1936) was the 'crowning achievement of pre-apartheid segregation' (King 2005: 66), as it expanded the native reserves system and the role of the tribal authorities. Rural people applying for land were granted a 'Permission to Occupy' (PTO), which was also called a 'Right to Occupy', to establish that the land had been allocated to them by their chief. However, PTOs were not recognised by financial institutions, as they were regarded as limiting investment opportunities, and PTO holders could be forcibly removed when the government deemed it necessary (King 2005).

> Colonialism and apartheid rule disrupted the traditional institutions of chieftainship and kinship:
>> In a context in which there were multiple institutions with a customary claim – such as gender institutions, age groups, clan assemblies, hereditary (customary) alongside bureaucratic (state-appointed) chiefs – colonial powers privileged a single institution, the bureaucratic chief, as the 'customary' authority, whose version of custom would henceforth be enforced as law. (Mamdani 1999: 98, in Oomen 2005: 18).

In this manner, traditional chieftaincies were no longer representative of traditional values, but became closely associated with colonial oppression. Ntsebeza (2000: 285) states that 'an African chief, as trustee of the community's land, may alienate land with the consent of the chief's council and without the direct participation of the community'. The right to control land allocation was, therefore, squarely vested in the chieftaincies. Levin and Mkhabela (1997) suggest that the institution of the chief as 'custodian of the land' was a colonial creation, produced by the need to create a customary land tenure system. Traditional communal tenure and perceptions of communalism, therefore, were myths developed during the early colonisation of southern Africa. These insights from history are important to bear in mind when examining the relations of access to water at the regional and local level, and they help to explain some of the phenomena observed.

Force and featherbed: The growth of mining and agriculture

A watershed event in South African history was the discovery of diamonds and gold in the 1860s, which propelled the country into the industrial age (Marais 2001; Sparks 2003b; Turton, Schultz et al. 2006). The burgeoning mining towns increased the demand for foodstuffs, which was an opportunity that African peasants sought to grasp. As Bundy (1988: 367–368) writes:
> After their dispossession following the *Difaqane*, blacks had been making something of a comeback, gradually working their way back onto the land as tenants and sharecroppers. A whole new peasant class emerged toward the end of the nineteenth century, and established a degree of independence.

Over 80 per cent of white-owned land (almost half a million hectares) was farmed by African tenants by 1860 (Hanekom 1998). African farmer enterprises proved to be efficient and competitive: they adopted new technologies and entered new industries, and came to represent an increasing threat to white dominance, who resorted to the argument that 'because of labour shortages, they could not compete with their African counterparts' (Hanekom 1998: 6).

At a stroke, the 1913 Land Act changed all this. The Act reserved only about eight per cent of farming land for Africans, rendering it illegal for them to engage in farming practices outside of these designated areas, and eliminated the potential threat of competition with white farmers. The ultimate aim of the Act was to create surplus labour for the mines and the white agricultural sector, effectively to stop the growth of an independent African peasantry and force them to become migrant workers (Bundy 1988). The gradual rise of an African peasantry was nipped in the bud.

Thus was laid the foundation for an exploitative labour-migration system at the heart of South Africa's wealth accumulation, which served to split the social fabric of families. Able-bodied men were recruited to work for a pittance in the mines, and there they stayed for weeks on end before returning to the reserved territories where their existence was permitted only when they were not labouring.

In the aftermath of the Anglo-Boer War, the Boers were also suffering. Returning to find their farmsteads burnt out, they flocked in their thousands to the cities to look for work. During the mining boom, increasing numbers of white *bywoners* (squatters) found themselves squeezed into the cities, where they lived 'cheek by jowl' with their equally poor African counterparts. Boers and Africans were members of two proletariats in competition for work (Sparks 2003b). The solution to the problem of poor whites was a combination of discriminatory legislation, which prevented Africans from owning farmland, but which provided assistance from the Land Bank for whites to acquire land for farming, and provided strong support services and subsidies to the struggling white farmers. This concoction of remedies to support the *bywoners* was to create 'forevermore a featherbedded and inefficient agricultural industry' (Sparks 2003b: 133), rendering the 'freedom-loving Boers…[as] welfare farmers, most of whom would not have survived without taxpayer largesse' (Bate & Tren 2002: 260).

Deepening dispossession: The National Party's segregated development

The coming to power of the National Party in 1948 and the accession to the office of Prime Minister of the possessive and extreme nationalist, DF Malan, marked the beginning of half a decade of systematic segregation. In power, one of the ruling party's preoccupations was to secure the political interests of the Afrikaners and to promote Afrikaner-led business initiatives, in the process instituting affirmative action which favoured Afrikaners over the British, both politically and commercially (Harrison 1981, in Bate & Tren 2002). The Broederbond, a secretive organisation which was open only to Afrikaner males, worked towards securing this aim. However, the apartheid government continued to depend on the British-initiated system of indirect rule, rather than inventing new institutions. The new turn of events heralded a far more brutal and oppressive regime, which peaked in the introduction of the deeply-hated Pass Laws and strict influx regulations.[5] The logic of apartheid was to use space to separate racial groups through a series of policies that culminated in the creation of so-called Bantustans, which were later called homelands[6] (Pickles & Woods 1992; Mamdani 1996; Terreblanche 2002; Sparks 2003b).

The Bantu Authorities Act (No. 68 of 1951) recognised tribal authorities as the main systems of governance, and, through the continuation of indirect rule, they became the primary lever of rural local government and played a central role in land allocation throughout the apartheid period and beyond. Ten Bantustans were created, and four of these – Transkei, Bophuthatswana, Venda and Ciskei – were eventually to become independent states and, as a result, their inhabitants were stripped of South African citizenship. The independent territory of Bophuthatswana even had its own water legislation (Thompson 2006). The others – KwaNdebele, Gazankulu, Lebowa, QwaQwa, KwaZulu and KaNgwane – remained self-governing territories, essentially creating labour reserves to supply the burgeoning mining industry with cheap migrant labour (May 2000).

A legacy of inequality

Apartheid served to entrench a socio-spatial division of the peoples of South Africa, with marginal portions of the country set aside for the indigenous African peoples in the homelands. The process of creating these homelands entailed the forced uprooting of more than 3.5 million people, with the splitting of families and clans and the rending apart of the social fabric. The system of indirect rule employed to govern the homelands meant that chiefs not amenable to the government were removed or sent to re-education camps.

In terms of resources, the spatial apartheid wrought through the creation of Bantustans is still entrenched in today's society, with the major part of fertile agricultural land still in white hands. This reflects the extent of overcrowding and land deprivation which resulted from the apartheid policy. Oppressed, discriminated against and stripped of their rights as citizens, the greatest symbol of the South African peoples' suffering under colonial and apartheid rule is the staggeringly skewed land distribution. 'The extent of land dispossession in colonial and apartheid South Africa dwarfs that of other southern African states' (Hall 2004: 219).

This huge inequality in distribution also affected access to water resources. According to Cullis and Van Koppen (2007), some 95 per cent of water for agricultural purposes was controlled by white farmers in the Olifants River Basin, and they speculate that this figure is probably representative of the situation in other parts of the country as well. The government also forcibly relocated entire communities to arid rural locations with poor soils and limited water resources, both to clear out unwanted shanty towns and to make room for commercial forestry, agriculture and National Parks on premium land.

The skewed distribution of resources was augmented by scant social spending. In the early 1990s, the State still spent three times more on social programmes for whites (education, health, social pensions and housing) than for Africans (Van der Berg 1998). The disastrous Bantu education policy adhered to under apartheid spent, on average, a miserly $25 per annum on each African pupil, as opposed to $150 per annum on each white pupil (Kenney 1980: 119, cited in Sparks 2003b: 196). Many African teachers preferred to resign rather than spend time in a system intent on 'bending young minds to an acceptance of inferiority' (Sparks 2003b). Such unequal educational opportunities have long-lasting repercussions, as they transmit across generations and affect the possibilities of employment, thereby cementing racial inequalities in the labour market (Van der Berg 1998).

The evolution of water rights regimes

Having outlined the broad strands of relevant South African history, it is now time to sketch out how water-use rights regimes evolved in this context, as this will be crucial in order to understand the dynamics of the later reform processes. This section provides an overview of the historical trends, highlighting how water came to be linked to land possession, and also describes how a hybrid system of riparian rights and permits developed prior to the promulgation of the NWA.

Roman-Dutch law comes to the Cape

Whilst land was regarded by the first Dutch settlers as freehold, water resources were a different matter. The Dutch, accustomed to plentiful water in their country of origin, and drawing upon Roman legal traditions, treated flowing water as *res communes*. This implied that all members of society could make use of the resource. In other words, water belonged to no one in particular, but to everyone in common. However, only rivers with a strong, sustained flow were classified as *res communes*. Rivers with variable flows could be labelled private *res in commercio*, which essentially meant that water could be owned by whoever abstracted it.[7]

A cornerstone of the Roman law was the doctrine that implied that a landowner was also the owner of everything above and beneath that piece of land.[8] This included groundwater, provided it was not the source of flowing or running water. The State was conceived of as *dominus fluminus* (literally meaning having 'dominion over flows') and intervened when competition for the use of water – for navigation, fishing or other purposes – intensified. Intervention was mainly through the issuing of praetorian edicts,[9] or through other forms of control such as servitudes, interdicts and taxes. When peaceful resolution was not forthcoming, intervention by the praetors was on an *ad hoc* basis (Burger 2006; Thompson 2006).

The form of Roman-Dutch law[10] that evolved through the practice of the early South African jurists maintained the distinction between public and private waters as set out by Roman judicial practice, but was rather more muddled with respect to the position of the State vis-à-vis water. Thompson notes that 'it was uncertain to whom the water belonged' (Thompson 2006: 27). Some jurists held that it belonged to the citizens in common property, whereas others interpreted the concept of *dominus fluminus* (or *dominus fluminis* – Thompson uses these terms interchangeably) to mean that the State was indeed the 'owner of the river' and had proprietary rights. Though the concept is a fuzzy one, Thompson argues that the general deduction is that the common use rights of all citizens were recognised, whilst the State maintained powers of regulation, akin to those practised in Roman times, through *ad hoc* edicts.

The British doctrine of riparian rights

With the arrival of the British, things gradually began to change. The mercantile power of the Dutch waned and the British moved in to fill the vacuum, gaining ever greater influence. Thus it was that in 1820, the English doctrine of riparian rights, which basically granted water-use rights in accordance with land ownership, replaced the Roman-Dutch practices of *dominus fluminus*. Riparian means adjacent to a river, and the principle of riparian rights essentially entails that all holders of property that border on a flowing water source are entitled to the use of the flow, irrespective of the time of their first appropriation. Under English common law, State control of water was an alien concept, and the Crown had no authority over water in rivers, except for navigation rights. Hence, the water in the river generally belonged to owners of riparian land (that is, on the riverbank), with an equal share for all owners (Getzler 2004; Burger 2006). Riparian rights were an integral part of land rights, as opposed to subsidiary land tenure rights, such as an easement or servitude (Hodgson 2004; Hodgson 2006).

Along with the introduction of the riparian principle came the idea of reasonable use, which was derived from the English common law tradition. Essentially, reasonable use entailed that each riparian user was allowed to use water in its natural flow as long as enjoyment of the flow was also secured for other users, and that interference with any particular user's rights by the Water Courts was allowed in order to protect the equal rights of all users along a river. Thus a system of common property rights along specific stretches of river was established, where groups of individuals voluntarily joined together to engage in decision-making on the allocation and exercise of rights. The riparian system was thus representative of a common-property system (Backeberg 1995; Bate & Tren 2002; Tisdell 2003; Getzler 2004). However, where the Roman-Dutch principle of *res communes omnium* essentially stated that water resources were open to all in society, the riparian principle limited access to the common-pool resource to those with rights to land, rendering it a closed commons.

The capricious creation of colonial judicial policy

Over time, the practice of judicial policy completely eroded the role of the State as *dominus fluminus* (Thompson 2006: 68), ensconcing the doctrine of riparian rights in judicial practice. The riparian principle was rather ill-suited to South African climes, however. Developed as it was in the 'damp and water-rich' climate of England, its evolution had been dictated in the main by disputes over flows to power water-mills rather than water abstractions for consumptive use (Rose 1994a).

In the United States of America, Angell's *Treatise on the law of watercourses*, published in 1824, sketched out several key concepts relevant to the practice of water use for agriculture in the arid western USA (see, for example, Getzler 2004: 273). Of particular importance was the notion of rights through 'prior appropriation', which adhered to a first-in-time, first-in-right logic. This publication was used as a guide by the South African Judge Bell in several rulings, but evidently he had only read parts of it, or, alternatively, he chose to ignore these sections. Several authors point to the fact that South African history might have turned out rather differently had Judge Bell indeed taken account of the sections dealing with prior appropriation (Lewis 1934; Burger 2006). In Lewis' words:

> One wonders whether Judge Bell's views would not have been very considerably modified and whether the law [in South Africa] would not have taken a very different course had he known of the later experience of the semi-arid Western States of USA, where irrigation is now chiefly practised, and of Italy, the home of the Roman Law, in both of which countries the principle of priority of time has almost completely swept away the old doctrine of reasonable or proportional use.
> (Lewis 1933: 42, cited in Bate & Tren 2002: 69)

Hall claims that this decision cannot be overstated, as it laid the foundations for the South African common law with respect to water resources (1939, quoted in Bate & Tren 2002). Burger concedes this point, stating: 'It is indeed a great pity that the learned judge ignored the principle contained in Dig.43.20.1. Had he done so, South Africa would have been spared 150 years of riparian doctrine and the difficulties associated with its implementation' (2006: 43).

The failure to acknowledge prior appropriation by the judicial courts as an alternative approach to water resource allocation implied that the riparian doctrine came to dominate thinking in South African water rights practice. Bate and Tren (2002) argue that irrigators would have preferred the doctrine of prior appropriation, as it would have reduced uncertainty. It is very difficult to tell what the impact would have been if the prior appropriation doctrine had been instituted. On the one hand, it is likely that it would have reduced the uncertainties for farmers, but on the other hand, it would have made it very difficult for newer arrivals, such as industry, to gain access to water.

What the retention of the riparian principle did do, though, was to intimately link water and land rights. Thus, although the author of the National Water Resources Strategy (NWRS) argued that the water legislation in place prior to the 1998 NWA (such as the 1912 Irrigation Act and the 1956 Water Act) were not discriminatory, the 1913 Land Act certainly was, and, by linking water access to land through the recognition of the riparian principle, by implication, so were the 1912 and 1956 Acts (interview, 5 May 2006).

The next section will describe some of the core elements of the 1956 Water Act, in order to appreciate the nature of the transition to the NWA of 1998.

The Water Act (No. 54 of 1956)

The Irrigation and Conservation of Water Act (No. 8 of 1912) basically consolidated the common law tradition, and gave preference to agricultural water use over industrial use. One significant change was that perennial as well as intermittent rivers were declared to be public, and the concepts of normal and surplus flows – aimed to make the most of floods of intermittent rivers – were introduced. The next four decades saw the expansion of industry, with a particularly rapid growth in mining (Bundy 1988; Marais 2001), and the consequent increase in demand for water led to increasing dissatisfaction with the 1912 Irrigation Act. A comprehensive evaluation of the legislation by the Tomlinson Commission, which was appointed in 1950, was the key trigger that led to the enactment of the 1956 Water Act.

A key feature of this Act was that, whilst retaining the principle of riparian rights, it facilitated the extension of government control by enabling the Minister of the newly established Water Affairs to declare so-called Government Water Control Areas (GWCAs), and thus to regulate water use through the issuing of permits or quotas, attenuating existing property rights (Bate & Tren 2002; Backeberg 2005).

The distinction between normal and surplus flows was retained. Normal flow was defined as the visible and flowing water in a stream that could be used without storage, and referred not so much to a quantity of water as a flow tempo (Water Act, No. 54 of 1956, section 9A). Proportional use – the 'fair share' – of this water to all riparian users was the norm. Section 9B in the Act allowed users to pump additional water over and above their fair share in periods of high or 'surplus' flow. In the opinion of one of the senior advisers in the Department of Water Affairs and Forestry's Directorate: National Water Resource Planning, the 'surplus use' resulted in farmers pumping for the equivalent of two months too long (interview, 25 April 2006).

In terms of trading, it was prohibited to transfer quotas or permits, but two contiguous riparian owners could freely deal with their allocated reasonable share of riparian rights, although where land was not contiguous, seepage or return flows had to be taken into consideration. Transfer of rights from riparian to non-riparian owners was also prohibited. However, a drought in the 1980s prompted illicit trade in quotas on the Crocodile and Orange rivers (Armitage 1999).

The story of the evolution of water rights regimes must be seen against the backdrop of the regime of dispossession and segregation that followed in the wake of colonial expansion. Whilst the 1956 Water Act facilitated increased government control, it still adhered to the principle that water access should be defined through land acquisition. The quotas, or permits, issued by the government were a means whereby riparian rights were attenuated, but the quotas were still largely dependent on riparian ownership of land. The non-tradable nature of the quotas ensured that the government had substantial control in the GWCAs. Outside of these areas, it was primarily the riparian principle that was the norm.

Transition

The water reform efforts were made possible through the transition to democracy and the coming to power of the ANC, which was transformed from a liberation movement to a party in power. It is necessary to understand the larger shifts in macroeconomic policies that occurred around this time in order to appreciate the climate in which the water reform took shape. The section that follows describes some of the key aspects of the transition, before turning to discuss the main drivers of reform.

A fledgling democracy: Crafting policies for change

The ban on democratic movements was lifted in February 1990, discriminatory legislation was quashed and constitutional negotiations began (Terreblanche 2002). This transformation ended 'the most systematic programme of racial segregation and oppression that the modern world has seen' (Sparks 2003b: vii). The ANC, the South African national liberation movement, had been founded in 1912 to promote the rights of black people. The organisation was banned by the National Party in 1960, and in 1961 the military wing of the ANC, Umkhonto we Sizwe (Spear of the Nation), was established with Nelson Mandela as its head. Three years later, Mandela, along with a handful of fellow ANC members, was sentenced to life imprisonment for sabotage at the infamous Rivonia Trials. During the 1980s, the ANC became the symbolic centre of resistance against the repressive apartheid regime by mobilising civil society and engaging in guerrilla action against the government (Marais 2001). With the dawn of democracy, the organisation faced a radical transformation from being a liberation movement to populating the corridors of power (Jordan 2004). On 10 May 1994, Nelson Mandela was inaugurated as President of the first democratically elected government in South Africa, marking the transformation from an apartheid state into a non-racial society. The fledgling government's focus was unwaveringly on the need to redress past inequities and work began in earnest on reforming old laws (Nattrass 1994; Marais 2001; Sparks 2003a).

The ANC government faced staggering levels of inequality, poor educational policies that had left most blacks uneducated, unskilled and without formal jobs, and a weak civil society, all of which demanded radical thinking on policies that would be more suitable for the nascent 'Rainbow Nation'. Although the birth of the new democracy had been a remarkably bloodless one – nothing less than a miracle – there was a long haul ahead to nurture it to maturity.

The challenge for the ANC was to 'come up with a consistent set of policies which promoted economic growth and job creation, alleviated white fears and boosted business confidence – while at the same time supporting redistribution, affirmative action, small business development and trade union demands' (Nattrass 1994: 346). This was a tall order by any standards. As a liberation movement and a 'broad church' that was at pains not to alienate its grassroots supporters, the ANC had little practical experience of staking out economic policies. Now it faced the challenge of effecting a gradual change from a highly protected isolationist siege economy into a competitive player in the global marketplace (Terreblanche 2002).

The ANC's election manifesto, the Reconstruction and Development Programme (RDP), was a piece of development policy with socialist resonance that was aimed at redressing past inequities through socio-economic and institutional reform, including educational and cultural programmes, employment generation and human resources development (Villa-Vicencio & Ngesi 2003). Key to the RDP was the emphasis on basic service provision and the view that the State needed to be restructured in order to facilitate a more equitable distribution of resources and deal with the socio-spatial distortions of the apartheid era (see also Bond & Khoza 1999; Maharaj & Ramutsindela 2002).

However, this initial socially-oriented policy was relatively quickly subsumed by the Growth, Employment and Redistribution policy (GEAR). This was a macroeconomic strategy that emphasised liberalism, deregulation and giving a loose rein to market forces with a concomitantly reduced role for the government (Villa-Vicencio & Ngesi 2003), and was 'blatantly misrepresented as the concrete form of the ditched and more ambitious and progressive RDP' (Fine 2003: 571).

This marked turn in policy generated considerable controversy, and presented a government already facing budgetary constraints and social pressures with the major dilemma of reconciling a 'social, rights-based, gap-filling and developmental approach with an approach based on productivity and efficiency' (Perret 2002). For example, the deregulation in agriculture signalled the end of a long period of State intervention in a strategically important sector, and caused a shift towards more high-value crops. While Vink and Kirsten (2000) argue that deregulation potentially results in a net welfare benefit, Hall (2004) argues that the process causes the gap to increase between 'winners' and 'losers' and that the State's emphasis on creating a cadre of commercial black farmers 'sits uneasily with its removal of subsidies and other support which have combined to produce a uniquely hostile environment for new entrants into agriculture' (Hall 2004: 220).

The nature of the negotiated settlement: Protecting property

How and why did this change in policy happen so fast? An organisation that played such a pivotal role in the anti-apartheid struggle would be expected not to have succumbed so easily to the flawed logic of the Washington consensus[11]. The irony, according to Fine, was that the ANC 'discovered neo-liberalism just as it was at its most extreme *and* its most vulnerable, in light of theoretical, empirical and policy failings' (Fine 2003: 572, emphasis in original). It is ironic that the political capital built up by the ANC during the liberation struggle is now being spent to enforce neoliberal structural reforms (see also Bond 2004; 2000; Carmody 2002). According to Carmody,

> the negotiated nature of the settlement meant the basic maintenance of the previous economic system, including respect for private property 'rights'. Thus, rather than enforcing redistribution and resource mobilisation internally, 'industrialisation by invitation' – drawing capital from overseas – became an attractive strategy for the South African state.
> (2002: 260, footnote omitted)

Thus, although the new Constitution (Government of South Africa 1996) was lauded for leading on social issues, such as recognising access to drinking water as a human right, it also protected existing property rights through section 25, or what became known as the 'property clause', in the interests of economic growth.

The Constitution's property clause and the emphasis on economic efficiency were reflected in the approach to land reform. Land being a hot issue, land reform was seen as vital to facilitate a smooth transition, and appreciating the workings of the 'land question' is also crucial in order to understand the dynamics of water reform. Land reform was a constitutional mandate, which rested on three pillars: restitution,[12] redistribution[13] and tenure reform. The protection of existing property through section 25 of the Constitution, combined with the government's concern with economic growth, resulted in land redistribution processes coming to adopt a willing seller, willing buyer approach to land reform. Although many have attributed this approach to the influence of World Bank policies, Lahiff (2005) argues that it was, in fact, mainly the government's own focus on economic prudence that underpinned the market-based logic to land redistribution. The ambitious goal of redistributing 30 per cent of land within the first years of democratic governance quickly resulted in disappointment, and a marked shift in policy in 2001 – signalling the transition from the Mandela years to the Mbeki years – saw the objective being reformulated as 30 per cent redistributed by 2015.

The land reform also brought to the fore issues of traditional authority. It underscored the rural-urban divide, but also went to the core of thorny questions about democracy versus traditional structures of governance. The next section deals with this re-emergence.

The resurgence of traditional authority

> *In the new South Africa, chiefs will melt away like ice in the sun.*
> (Civic leader, Gazankulu, quoted in Levin & Mkhabela 1997: 153)

Levin and Solomon (1997), amongst others, have emphasised the urban bias of the ANC's liberation struggle. This did not stem so much from the absence of rural resistance – although this did certainly occur – but from the fact that the ANC's tactics and strategies in the struggle for liberation were mostly drawn from 'an emergent petty-bourgeoisie and the ranks of the proletariat in the urban areas' (Levin & Solomon 1997: 175).

Whilst the negotiated settlement saw the ANC placing increasing emphasis on the need for economic reform, the party did not want to alienate traditional leaders as it foresaw the possible coalition of traditional leaders as a potential opposition force (Cliffe 2004). Traditional leaders were, therefore, able to assert considerable influence on the new Constitution, through a clause that allowed them to serve *ex officio* in local government, as well as to wield power via the establishment of Provincial Houses of Traditional Leaders and a National Council of Traditional Leaders (Cliffe 2004; Oomen 2005).

These powers were further strengthened by the passing of two important Acts, the Traditional Leadership and Governance Framework Act (No. 41 of 2003) (the TLGFA), and the Communal Land Rights Act (No. 11 of 2004) (CLaRA). These acts have been surrounded by a lot of controversy (Claassens & Cousins 2008), and the CLaRA was legally challenged by four communities.[14] The TLGFA and CLaRA were regarded as sister acts, which empowered the chiefs to a great extent. In Cliffe's (2004: 355) words:

> Tribal authorities were apartheid creations designed to bolster that system. They sparked rural rebellions and mass arrests throughout South Africa. The TLGFA gives them perpetual life. The Communal Land Rights Act (CLaRA) goes on to give them unprecedented powers over communal land that surpasses any that they previously enjoyed.

CLaRA basically provided for the transfer of title from the State to communities, as part of the tenure upgrading scheme. By 'communities', the Act effectively meant existing traditional authorities, and these authorities and the chiefs were thus granted discretionary powers over land allocation. This disregarded the family-based nature of land rights in rural areas. Instead, it provided that rights were vested in 'communities' or 'persons' (Claassens 2003).

A brief review of the political economy context

Black economic empowerment and corporate interests

In the RDP, the notion of black economic empowerment (BEE) was initially conceived of as a means to facilitate the redistribution of productive resources to those groups that had been oppressed and disadvantaged under the apartheid regime. However, the notion of BEE evolved over time into a process of affirmative action that provided black individuals, rather than groups, enhanced opportunities (such as through the preferential granting of shares). It became more concerned with how black people could access the returns of higher economic growth rates than with real redistribution of productive assets (Ponte, Roberts et al. 2007). This rendered the BEE initiative vulnerable to criticism that it was enriching a small black élite. Such criticism prompted the government to launch a second phase[15] of BEE, now dubbed 'broad-based' black economic empowerment (BBBEE), to emphasise the intended broadly inclusive approach and to make its policies more palatable to the ranks of the Congress of South African Trade Unions (COSATU) and the South African Communist Party (SACP). Nevertheless, some observers still hold that the BBBEE is an attempt to legitimise the government's neoliberal policies through facilitating the establishment of a black capitalist class, which in turn guarantees the survival of the white capitalists and their property rights (see, for example, Malikane & Ndletyana 2006, cited in Ponte et al. 2007: 934). There are also fears that such a route to 'empowerment' will nurture a culture of cronyism and nepotism, generated through the close linkages that exist between ANC politicians and individuals in business (Southall 2004). Others argue that BBBEE is a necessary move towards 'facilitating the socio-economic functioning of a society that, for historical reasons, would otherwise be doomed to large-scale civil strife' (Iheduru 2004 & Klasen 2002, cited in Ponte, Roberts et al. 2007: 934), and that fostering a black capitalist class would provide a feasible route to achieving the much vaunted 'trickle-down effect' of economic growth. Hence, the BBBEE approach contains contradictory objectives, and its implementation has not been subjected to academic analysis to any considerable extent. Corporations often interpret BBBEE charters in purely managerial and technical terms, and are therefore vulnerable to charges that they implement BBBEE merely to provide a route to acquiring legitimacy (Hamann, Khagram et al. 2008). In a sense, then, the process of BBBEE has moved from a political terrain, where redistribution is at least theoretically possible, to a managerial terrain, where discussions tend to be technical and codified (Hamann, Khagram et al. 2008).

The most high-profile attempt by the government to promote BBBEE occurred in the mining sector (Hamann, Khagram et al. 2008). As noted earlier, mining was a pillar of the South African accumulation strategy, and it accounted for the greatest share of export earnings (Marais 2001). Powerful multinational corporations, such as De Beers and Anglo American, dominate the mining industry, and given the emphasis on globalisation and integration with international markets that became evident in ANC economic policy shortly after transition, these companies at the time enjoyed a great deal of leverage in terms of influencing economic policies (Marais 2001). The so-called Mining Charter (Republic of South Africa 2002) proclaimed a commitment on behalf of mining companies to achieve at least 26 per cent black ownership of the industry's assets within a 10-year period. However, the newspaper *The Sowetan* reported that

> the mining houses still call the shots…government has been very slow in pushing the transformation of the industry thanks to the need to bend over backwards and accommodate the concerns of the industry.
> (Editorial, *The Sowetan* 16 August 2002: 11, cited in Hamann, Khagram et al. 2008: 21)

Big business was increasingly starting to perceive BBBEE as a risk that threatened investor confidence – investors whose reactions Hamann et al. (2008: 30) describe as '...swift, relatively ignorant of the broader decision-making processes, and perhaps unduly influenced by information sound bites'. In response, many South African multinationals (De Beers and Anglo American included) relocated their headquarters overseas, thereby removing their major assets from the reach of the South African state. Thus,

> far from being part of building a 'developmental state', the complexity of the provisions developed under codes and charters, coupled with low state capacity to engage meaningfully in the details of these provisions, means that 'state control' is increasingly exercised via 'outsourcing' of the development, monitoring and auditing of BEE-related provisions to an emergent industry of consultants and auditors. In this way, there is a diffusion of responsibility and the process of policy-making is depoliticised to a certain extent.
> (Ponte, Roberts et al. 2007: 949)

Commercialisation, 'de-agrarianisation' and the notion of 'two economies'

The commercial farming sector contrasted with the mining sector in that its political clout stood in stark disproportion to its economic importance. Agricultural capital was mainly in white hands, and white commercial agriculture had traditionally been a vital political constituency of the National Party governments under apartheid (Bate & Tren 2002), and had built up a strong agricultural union, Agri South Africa (AgriSA), to represent its interests. The black counterpart to AgriSA, the National African Farmers' Union (NAFU), is neither a particularly large nor powerful interest group, but Hall (2004: 224) holds that it does provide a 'ready partner for a programme of deracialisation in the commercial sector'. In 2001, the Strategic Plan for Agriculture was launched to provide a platform for the future partnership of AgriSA and NAFU and to guide government policy. In the words of Hall:

> The alliance that has emerged through this process involves, in the first place, white commercial farmers and agricultural capital, which has an interest to maintain property prices and confidence in the land market, and to acquire black neighbours engaged in the same forms of production – including for the purposes of protecting access to international markets. Secondly, it involves government, whose interests lie in revenue and stability, and therefore the growth of the commercial sector alongside its deracialisation, and the growth of a black middle class, not least as a political buffer. Finally, and possibly most marginally, it involves black 'emerging' commercial farmers, whose interests are to gain access to state resources and become beneficiaries of the deracialisation of the sector and related BEE initiatives.
> (Hall 2004: 224)

The change of gears in macroeconomic policy formulation rendered the pathway of blacks engaging in 'emerging' commercialisation a narrow one. Whilst land reform provided crucial resources to a few, the general deepening of capitalist relations meant that transferring assets to the poor was 'anomalous, since they lack the means to engage in capital accumulation in commercial agriculture' (Hall 2004: 225). Concomitant with these processes, evidence is accumulating that the levels of inequality in the countryside are increasing (Du Toit & Neves 2007), and that a process of what is termed 'de-agrarianisation' is occurring (Bryceson 1996, cited in Du Toit & Neves 2007: 154), which implies a slow collapse of agricultural production in the homelands.

A popular way to describe South Africa's socio-economic condition is through using the metaphor of 'two economies' (Bond 2007; Cousins 2007; Du Toit & Neves 2007). This metaphor construes the South African situation as comprising two isolated economic realms that exist in disconnected parallel: the realm of 'traditionalist' modes of production and sustenance, and the realm of modernised, industrialised production. In agricultural terms, subsistence agricultural production is pitted against the modern agricultural sector, where modernising is equal to the consolidation and commercialisation of farming activities. The gist of the 'two economies' discourse is that those 'stuck in the backwaters' need to be given a route of access into the 'first economy'. However, as Du Toit and Neves are at pains to show, such a disconnected view is uncalled for. Rather than attempting to eliminate the 'second economy', they argue, it would be better to find measures that could mitigate existing power imbalances with the aim of reducing inequality. Moreover, instead of construing symbolic myths, policy-makers would be better advised to explore what is really there, on the ground, instead of basing their strategies on simplistic assumptions. This calls for a 'less pathologising, less normative gaze' (Du Toit & Neves 2007: 171). It also requires abandoning the naïve belief that integration will automatically bring benefits for those on the outside. Rather, policies that 'value and support the fragile survival strategies that take shape on this hostile and difficult terrain' are needed (Du Toit & Neves 2007: 171).

The idea of poor people being trapped in structurally disconnected productive spheres is associated with another discourse, which focuses on poverty arising out of a lack of an adequate level of basic service delivery. Such a conception situates poor South Africans first and foremost as passive citizen-consumers. Poverty reduction, therefore, is conceived of essentially as a matter of supplying such services, particularly to rural remote areas, 'independently of whether these services can be used, afforded or are even locally necessary' (Bank et al. 2006, cited in Du Toit & Neves: 50).

The above shows how different currents are at work in the South African context, and how there is a tension between divergent understandings of cultures of individuated property and customary rights, and how the particular political economy context contributes to shaping the political terrain. The next section considers how water rights reform fits into these debates.

Getting the Act together: Processes and drivers

The geophysical backdrop

Before going on to describe the main drivers of reform, it is necessary at this point to provide the reader with an impression of the geophysical characteristics of South Africa's water situation. South Africa is routinely characterised as a semi-arid country. It depends in the main on surface water resources for most of its urban, industrial and irrigation requirements, with a mean natural – that is, undeveloped – run-off estimated to be roughly 49 billion m³ per annum. Surface water resources have been intensively developed: about 320 dams of more than 1 million m³ of capacity have been built, which together make up more than 66 per cent of the annual run-off. According to the NWRS (DWAF 2004), the northern regions of the country are almost fully developed, whereas the south-eastern parts are quite well watered. Large porous aquifers occur in only a few places, and groundwater is therefore of apparently limited importance, though it is extensively used in rural and arid areas.

South Africa has four major rivers (which it shares with neighbouring countries) namely the Orange, the Limpopo, the Inkomati and the Pongola rivers. Irrigation use represents more than 60 per cent of total water requirements, urban requirements total about 23 per cent, and the other major sectors – bulk industrial, mining, rural and afforestation – make up the remaining 15 per cent. As stated in the section on the legacy of inequality, these resources are very unevenly distributed. Studies from the Olifants Water Management Area demonstrate that more than 95 per cent of water resources in that basin are used by the white minority, which provides an indication of the extent of inequality. Having briefly sketched the backdrop, it is now time to discuss the main drivers of reform.

The initiation of reform

Law cannot be understood outside social and political relationships. Whose ideas get instituted in law, and whose versions and visions get perpetuated, accepted and institutionalised, depends on power relations and the wider political context.

The huge inequalities left by the apartheid legacy were a key concern to the ANC, and one of the first areas to receive attention from the ANC government was the staggering gap in services. All-white suburbs accounted for more than 50 per cent of residential water use, and only 27–28 per cent of black households had running water, which was in striking comparison to 96 per cent of white households (Francis 2005). Since water is such a basic necessity, the ANC perceived water reform to be a top priority on its agenda, as is evident in the emphasis on water in the RDP (Bond & Khoza 1999), and the fact that the 1956 Water Act was one of the first to be subjected to reform. The crafting of the new water legislation was done with financial backing and help from the UK's Department for International Development (DFID).

The first Minister to be elected as head of the Department of Water Affairs and Forestry (DWAF) under the new democratic dispensation, was the human-rights lawyer Kader Asmal, who is of Indian descent. Hitherto, DWAF's ministers had been white farmers, and Asmal's appointment to this position thus represented a break with a century-old tradition. He served as Minister of DWAF from 1994 to 1999,[16] and from 1998 until 2000 he also served as Chair of the World Commission on Dams. Asmal's background in human rights law was influential in terms of coining the term 'Reserve'. The Reserve referred to the notion that a minimum quantity of water should be retained in the system to meet basic human and ecological needs. It was a ground-breaking concept which earned much publicity and which stood in stark contrast to the situation prior to 1994 when the DWAF had no jurisdiction over the homelands. It did not then consider including the provision of water to all citizens as part of its mandate, which Turton and Henwood (2002: 45) consider to be a 'manifestation of the institutionalisation of resource capture that had occurred throughout most of the twentieth century'.

Asmal, a charismatic figure with a strong personality, thus set about the task of closing the gap in services with characteristic zeal, and acted as a 'policy entrepreneur' (Cobb & Elder 1979), in the sense that he pushed the water policy process in a certain direction at an opportune time. During a conversation at the National Water Summit, one DWAF employee described Asmal as being 'arrogant... but you could speak to him – he would listen' (personal communication, 5 May 2006). His status as an internationally-renowned human rights lawyer made a profound mark on the formulation of the Act.

Several observers have commented that the emphasis on closing the backlog of services owed much to Asmal, but it was the next minister in line, Ronnie Kasrils, who was the one to take issue with the question of cost, and who introduced the Free Basic Services policy.[17] According to Turton and Meissner, quoting a personal communication with Bill Rowlston of DWAF, Asmal's primary aim was to address the social concern of the imbalances in water services provision, and 'it took some time before Prof. Asmal began to grasp the connection between water services provision and resource management' (Turton & Meissner 2002: 59).

Shortly after becoming minister, Asmal appointed Len Abrams and Barbara Schreiner (DWAF's Chief Director: Water Use and Conservation) as his advisers, and in 1995, the publication *You and your water rights*, drafted by Len Abrams, was published. Abrams, a water engineer by training, was also author of the Water and Sanitation White Paper. In April 1995, the Water Law Review Panel was appointed, with a legal and scientific portfolio, and included representation from the water use sector, rural communities, departmental officials and environmentalists (Garduño & Hinsch 2005; De Coning 2006).

Prior to the drafting of the National Water Policy, which was adopted as a White Paper by Cabinet on 30 November 1997, a set of fundamental principles for water law – the Water Law Principles (see Appendix B) – was produced by the Water Law Review Panel. The Water Law Principles drew up a framework for managing water resources, crafted in accordance with the concerns of the new democratic government, and focused on equitable and sustainable social and economic development.

Christo de Coning, an academic based at the University of the Witwatersrand and interested in public policy research, observed that Asmal's background as a lawyer was 'hugely influential' in terms of shaping the Water Law Principles. According to De Coning, the process was a legal rather than a policy process, and the issue of problem structuring was not given enough attention:

> It was defined that water should cease to be a private good, and be developed as a public good, and the challenge was to find ways of managing it as a public good. So this took much attention, rather than focusing on what were actually the problems in water.
> (Telephone interview 13 June 2006).[18]

The DWAF was heavily criticised for not producing any written drafts of the White Paper before the final version was released. However, the National Water Bill was published on 27 January 1998, and Parliament subsequently approved it as the National Water Act No. 36 on 20 August 1998.

During the drafting process, the drafters consulted widely in terms of international experience and were clearly influenced by IWRM principles (Brown & Woodhouse 2004; Garduño & Hinsch 2005). A workshop was organised in Pretoria in 1997 to present case studies and lessons from countries elsewhere, such as Mexico, France, Australia, Britain, Malaysia, India and even neighbouring Zimbabwe. Australia, in particular, constituted one of the most important sources of influence during the drafting process, as it was argued that it was very similar to South Africa in geophysical terms (Hadley Kavin, member of the drafting team, interview 10 August 2006). The model of Catchment Management Agencies (CMAs) was drawn from Australia. A potential pitfall lies in the physical similarities of Australia and South Africa overshadowing the socio-economic differences, particularly with respect to the capacity to build up such institutions. This is clearly reflected in the fact that of the 19 proposed CMAs (which it was initially envisaged, rather optimistically, would take a couple of years to establish), only one – the Inkomati CMA – was fully up and running at the time of research (2006).[19] But what emerged as the main drivers of reform?

Focus on services

The framing of policy is not simply limited to the choice of policy questions, but also to the setting of agendas, the bounding of institutional remits, the prioritising of research, the inclusion of disciplines, recruitment of committees, and so on (Stirling 2005: 224). In terms of agenda setting, there was a sense that Asmal concentrated so much on the question of extending service that he did not really grasp the issues inherent in resource management until much later, as noted by Turton and Meissner above. Burger (2006: 9) also concurs with this view, arguing that Asmal did not really have a ready idea of resource management principles, but rather ended up 'extending the existing permit system'. In Burger's own words:

> In 1998, Minister Asmal took a courageous step in totally abandoning the riparian principle. Unfortunately, the Minister had no clear principle at hand at that time that could serve as a substitute. It was recognised that parts of the prior appropriation doctrine could help, but no expert consulted by the Minister had apparently proffered a sound overall principle. He, therefore, adopted and extended the existing permit system set out in Act 54 of 1956 and made it applicable not only to some users, but to *all* water users.
> (Burger 2006: 9, emphasis added)

There was no clear consensus on what path to follow in terms of adopting a new legislative framework, and the prominent water law expert Robyn Stein was quoted as having argued for the retention of the current legislation and retaining the doctrine of riparian rights in its basic form, but with the addition of necessary adjustments.

The focus on services was further reinforced with the publication of a memorandum arguing for the continued separation of service and resource management by Asmal's Director-General, Mike Muller. The memo, entitled *'Omissions of the principle which recognises the difference between a pipe and a river'*, pledges that the Constitution itself makes this distinction, so consequently two separate Acts were developed to deal with services and resources respectively: the Water Services Act (No. 108 of 1997) and the National Water Act (No. 36 of 1998).

Whereas the management of water resources is placed under the aegis of the national government, the responsibility for providing water services lies with local government. This led to a gap in the attitudes of DWAF staff as well. According to one employee, the staff dealing with services were jokingly referred to as the 'internal department', since there was little cross-sectional communication between the people dealing with service delivery and those who were working on resource management. He continued, '…they're not really talking to one another. It's like a wall in here' (interview, 26 April 2006). This view reflected a prevailing sense among many members of staff of 'working in silos', and also how the 'bounding of institutional remits' (Stirling 2005: 224) worked against the 'integrated' nature of the IWRM paradigm.

The emphasis on services gave results. South Africa has an impressive record in terms of dealing with the backlog of water services that remained after the transition. In 1994, an estimated 14 million people lacked access to clean water, and the Department set about correcting this at a rate of about a million additional people connected to supply services a year, which is a truly impressive achievement and widely commended (see Eales 2011). It is far less likely that similar progress will be seen in terms of resources, for, as one academic remarked, 'water services are much more politically appealing than water resources' (interview, 26 April 2006). It also highlights how IWRM principles were subsumed by the more politically appealing focus on services (see Jonker 2007 for details), and the problems separation created in determining what would constitute legal domestic use (Kahinda, Taigbenu et al. 2007: see also section on Schedule One, pages 37–38).

Environmental concerns

In contrast to other countries, in South Africa it was the overarching concern with social issues that acted as a trigger for reform, argue Turton and Meissner (2002). But there are other versions of the story, too, such as the one held by a legal practitioner in the Inkomati who specialised in water law and who had been involved in the drafting process. She stated that the drafting of the water law grew out of a concern about the lack of rights on behalf of the environment, and described the process as follows:

> It all started out with a research project in the Kruger National Park in 1990 [initiated by the Water Research Commission], to find out about the health of the ecosystem, and chart environmental indicators. We then wanted to find out what the rights of the ecosystem were, what kind of claims we could make in terms of water quantity and quality of rivers flowing into the park… did we have any rights on behalf of the environment at all?
> (Legal practitioner, Komatipoort interview, 7 June 2006)

During the Kruger research project, she ventured, she became obsessed with the Park. She used to work as a prosecutor, but now wanted to get into environmental law because of this new obsession. A professor at the Lowveld College of Agriculture convinced her that it would make sense to specialise in water law, which prompted her to ask why the current water law was not taking the environment into account. In her view,

> the 1956 Water Act was very much geared towards irrigation; it was basically taking care of irrigation...it only provided rights for irrigation, and disputes were only for irrigation. There was no such thing as 'the environment' in the old days...However, in the eighties and nineties, people were becoming more aware of the environment: environmental awareness rose. So you needed to have a law that took this into account. That was how the water law reform was initiated. It started with the question of how one should provide water for the environment...But after the transition in 1994, with the new government in place, the question became *But what about people?*
> (Legal practitioner, Komatipoort interview, 7 June 2006, my emphasis)

Later, the feeling that the process of drafting the law had been hijacked by the environmentalists was strengthened during an interview (10 August 2006) with Hadley Kavin, a lawyer and member of the drafting team. This view is corroborated by De Coning's analysis of the process leading up to the promulgation of the Act. The task teams involved in determining policy implementation options were biased towards environmental interests, in that the environmental task teams tended to consist of more members and had more capacity than the others (De Coning 2006).

Whilst services and environmental issues formed the main focus of the reform effort, the DWAF's stance on water for agriculture was much more ambiguous. During consultations with stakeholders, a concern was raised that the Act did not view irrigation as being of specific importance to national food security or food self-sufficiency, which was a natural reaction from the farmers' lobby groups such as AgriSA. However, the ambiguity did not only concern commercial interests, but applied just as much to smallholders. Robyn Stein, a lawyer and a prominent member of the water law drafting team (who also headed the Water Tribunal at its establishment) admitted to the fact that the drafters had not taken the smallholders' situation into account when hammering out the principles (Van Koppen, personal communication, 2006), which contributed to a prevailing ambiguity on the DWAF's part with respect to the role of agriculture.[20]

Crucially, the Water Law Principles and the Act were regarded as too vague on the issue of how to go about allocating water in actual practice. A policy was needed to put flesh on these principles, which became known as the Water Allocation Reform Programme. Although the Act itself was promulgated in 1998, the process of fleshing out the principles of allocation did not start until five years later in 2003, with the appointment of the Expert Panel, and it took almost three more years to develop a complete policy on allocation. But before going on to describe the process of fleshing out the allocation reform in principle and practice, the reader needs to become more familiar with the main contents and contested issues of the NWA of 1998.

The National Water Act 1998: Key features and debates

Forceful drivers for reform, then, were concern for the environment and addressing the backlog in services, and the alleged lack on Asmal's part of a clear idea of resource allocation issues, had an impact on the nature of the drafting of the Act.

In the following section, I describe and analyse some of the key features and debates surrounding the NWA. This will provide a background to understanding how key ideas were conceptualised, and how tensions in understanding would carry over into the water allocation reform process, giving rise to a struggle for different interpretations. At the national level, the NWRS is the overarching instrument for managing national water resources. Its purpose is to 'set out the strategies, objectives, plans, guidelines and procedures of the Minister and institutional arrangements relating to the protection, use, development, conservation, management and control of water resources' (DWAF 2004: 8). Essentially, the NWA can be argued to represent an expansion of the notion of GWCAs, but rather than merely covering specified regions, it was now to cover the whole country.

The partitioning of the country into 19 Water Management Areas (WMAs), based more or less on drainage regions and with each being governed by a CMA, was in the spirit of IWRM principles. The purpose of the CMAs was first and foremost outlined as 'co-ordinating and promoting public participation in water management' (Anderson 2005: 1; see also Brown 2005), though it was envisaged that these responsibilities could be expanded to include setting and collecting water use charges and issuing water use licences (Schreiner & Van Koppen 2002). Originally, the first handful of CMAs should have been up and running within the first couple of years after reform. However, at the time of research, only the Inkomati CMA had been completed, and even that was far from being fully operational. I turn now to discuss the key notions of water-use rights and the debates emerging around these concepts.

Recognised water uses, the Reserve and Resource-Directed Measures

The Act (chapter 4, part 1, section 21) recognised 11 water uses, namely
(a) taking water from a water resource;
(b) storing water;
(c) impeding or diverting the flow of water in a watercourse;
(d) engaging in a stream flow reduction activity contemplated in section 36;
(e) engaging in a controlled activity identified as such in section 37(1), or declared under section 38(1);
(f) discharging waste or water containing waste into a water resource through a pipe, canal, sewer, sea outfall or other conduit;
(g) disposing of waste in a manner which may detrimentally impact on a water resource;
(h) disposing in any manner of water that contains waste from, or which has been heated in, any industrial or power generation process;
(i) altering the bed, banks, course or characteristics of a watercourse;
(j) removing, discharging or disposing of water found underground if it is necessary for the efficient continuation of an activity or for the safety of people; and
(k) using water for recreational purposes.

'Engaging in a stream flow reduction activity' was a controversial conceptual innovation. The forestry sector contemptuously referred to it as a 'rainfall tax', and there were constant discussions and debates about whether other uses, such as dryland sugar-cane farming, should also be considered a stream flow reduction activity.

The Reserve was one of the most innovative ideas that emerged during the drafting of the NWA. Essentially, the Reserve consists of two components. The first component, the human-needs Reserve, is based on the idea that a certain quantity of water should remain in the system to provide for the basic human needs of all South African citizens. The right to a minimum amount of water for sustenance is provided for in the Constitution. The Reserve, however, only provides a right to a basic domestic minimum (see, for example, Pollard, Moriarty et al. 2002 on the challenges of making the Reserve operational).

The second component, the ecological Reserve, is meant to ensure a minimum quantity of water to maintain environmental sustainability. In the NWA, the whole of chapter 3 is devoted to protection of water resources and deals with the development of a classification system for water resources and resource quality objectives, determination of the Reserve and pollution prevention. The Water Research Commission played a key role in terms of research related to the ecological

Reserve, whereas conspicuously little was done in terms of research on water for basic human needs. The NGO AWARD, however, does research on, among other things, how to meet basic human needs in practice, mainly in the Mpumalanga and Limpopo provinces.

Hadley Kavin, the lawyer and member of the water law drafting team, emphasised the importance of the technical aspects, and that the 'hijacking' of the process by environmentalists had profound implications. Up until the promulgation of the Act, there were standards that provided a certain threshold for allowable discharges into specific rivers. With regard to licences, the proposal was that conditions pertaining to the class would be subject to each individual licence. The intention was then to apply regulations, to be able, for instance, to curb bad polluters by issuing a five-year licence permitting them to discharge only so much during that time, but after that point the licence would have to be brought into line with current regulations. 'But we are now 10 years down the line, and I haven't seen a draft of the regulations yet' (interview, 10 August 2006). He was not the only one echoing concerns about the feasibility of the ecological criteria. Researchers at the University of Pretoria who were involved in determining the Reserve in specific catchments argued that 'the process of determining [the] Reserve is very lengthy – it takes about six years' (interview, 14 March 2006).

Among drafters and some academics, there was thus a sense that the resource classification system was hampering progress by the insistence on detailed, technical specification and classification, because licences would only be issued once the Reserve had been determined in a particular area.

Water use categories

Chapter 4, section 21 of the National Water Act deals with 'permissible water uses', and section 22 spells out the four broad categories of use of quantities over and above the Reserve. These are Schedule I, general authorisations, existing lawful use, and licences.

Schedule I

Water use permissible under Schedule I encompasses 'reasonable' water use – mainly for domestic purposes – that does not need an authorisation. It includes water for reasonable domestic use; small gardening not for commercial purposes; grazing (but not for feedlots), that occurs 'within the grazing capacity of that land'; storage and use of run-off water from a roof; and engaging in water use in cases of emergency, such as fire-fighting or human consumption. This categorisation implies a clear distinction between domestic and productive uses and implements the strict separation outlined by Mike Muller, then the Director-General of DWAF. During the interview with Hadley Kavin (10 August 2006), he related how

> [the drafters] wanted to enforce strong water control. We wanted to incorporate strong water control but the technical people could not work it out…If you look at an earlier draft [of the Bill], they had figures for domestic use written into the Bill. Under Schedule I, they could use a certain amount, a certain volume…The technical people could not support it at the end of the day, so then we left it out altogether. If you look at draft four or five, it's in there. Today, funnily enough, we're actually using it as a basis. We actually use it as a basis, but at that time the technical people could not support it.

There was thus a pervasive concern with control, not only in terms of classifying ecological categories, but also with respect to the abstraction of relatively small quantities of water.

Although it is described as an entitlement in the Act, the concept of Schedule I does not confer any rights. It allows, but does not facilitate or guarantee. The Schedule I category thus comprises what Hodgson (2004) terms a *de minimis* right, that is, you are allowed to use, but your use is not protected.[21] That people have a right to use water under Schedule I is very different from an assurance that they have water security. They only have a right to what is there; if there is nothing, then they cannot claim water security.

Furthermore, Kahinda et al. (2007) argue that apparent Schedule I uses, such as rainwater harvesting initiatives, are illegal, strictly speaking, under the current legislation. This is because, although the NWA allows such harvesting under Schedule I, section 6(1) of the Water Services Act (No. 108 of 1997) states, on the contrary, that such uses require authorisation from the Water Services Authority.[22]

General authorisations

General authorisations refer to authorisations of water use in a specific geographical area, or for a specific group, that is deemed to have little impact on the resource. The legislation establishes a procedure to enable a responsible authority, after public consultation, to permit the use of water by publishing general authorisations in the *Government Gazette*. Such an authorisation may be restricted to a particular water resource, a particular category of persons, a defined geographical area or a period of time, and requires conformity with other relevant laws. General authorisations differ in one major respect from licences in that they are not transferable. General authorisations were included in the NWA because they allowed a more flexible approach, they promised a potential for lessening the administrative burden, and they were contemplated as being primarily 'used in areas that are not under significant water stress' (Stein 2006: 2180). However, as the reform process unfolded, general authorisations were redeployed more proactively as a means of ring-fencing water for the poor.

Existing lawful use

Another controversial issue was the idea of existing lawful use. The term refers to water use that took place during the two years immediately before the Act came into force, and which was authorised under any law in existence prior to the promulgation of the Act. During an interview with a senior member of the drafting team, he explained to me how the term existing lawful use had come into being: 'In terms of the 1956 Act and the 1912 Irrigation Act, they [the white farmers] were given rights to use water...[then] we came across this term, *existing lawful water use*' (interview, 10 August 2006, emphasis added).

In the NWA, existing lawful use is defined as part of provisions to deal with the transition mechanism from the riparian principle to an administered authorisation system. Thompson (2006) states that 'for practical reasons, the change could not occur immediately'. DWAF officials, however, explained the retention of existing lawful uses primarily in economic terms. The energetic and capable Director, Water Allocations, stated how, even though the existing users had greatly benefited from skewed landownership and the associated access to water through riparian rights, their uses were allowed to carry over 'because the economy depends on that kind of use' (interview, 31 October 2006). Although talking warmly of wanting to achieve social justice and lamenting the slow progress of the reform, his perception of the problem was framed mainly in macroeconomic terms. When asked about the possible alternatives to carrying over existing uses, he enquired, 'What exactly is the population in the Inkomati?' He then stated that, in the interests of equity, 'you would be compelled to give 97 per cent of the water to black people. What would that do to the economy? I'll leave that for you to speculate on.' This invocation of strict egalitarianism served to make my question seem ridiculous. He went on to draw a parallel with the Johannesburg Stock Exchange Limited (JSE), conjuring a situation in which the total wealth of the JSE was divided equally amongst all the inhabitants of South Africa. This was a clever rhetorical device, which underscored the perception of risk. By framing the issue of water allocation entirely within macroeconomic terms, a 'closure' is achieved (Smith & Stirling 2006).

The Chief Co-ordinating Officer (COO) of the DWAF, a prominent black lawyer who used to head the Legal Services section, maintained that the primary reason for retaining existing lawful uses was that they (the department) would be exposed to allegations of expropriation if they did not do so. He also voiced the perception of economic risk as playing a significant role in the decision, thus invoking the powerful idea of the 'two economies' (Cousins 2005; Bond 2007; Du Toit & Neves 2007). 'We have to be careful, to make sure that we don't affect the economy. The first economy and the second economy: the first one is driving the second economy' (interview, 20 July 2006). This use of the 'two economies' was a widespread, pervasive and powerful trope (Throgmorton 1993) that underscored the perception of risk.

Thus, rather than citing practical or capacity problems, the explanations given for introducing the concept of existing lawful uses were rooted in discourses on upholding the economy and avoiding charges of expropriation. In doing so, officials in DWAF followed the most basic strategy for garnering political support. Rather than through coercion or intimidation, this was by 'evocation of social and political interpretations that legitimate the desired course for action' (Fischer 2003: 55). Existing lawful uses, instead of being retained due to the lack of capacity by the Department to deal with them, were portrayed as a necessity to keep the country afloat.

A slightly different take on the retention of existing lawful uses was provided by a senior (white) member of the water law drafting team, who was a retired lawyer at the time of research. Rather than arguing the case for the carry-over from a utilitarian point of view, he expanded on how the apartheid state had, in fact, been beneficial in some respects, as it had provided infrastructure, thus facilitating development that would not otherwise have taken place. This was in line with Turton and Meissner's (2002) depiction of the State as being on a 'hydraulic mission', which argues the case for those on the receiving end of that expansionist zeal, so the existing users were regarded as part of 'the exercise of building a nation' and should have been accredited as such. In parallel, the retired lawyer asked, 'why should we give back water to the historically disadvantaged individuals when they only had dryland in the first place?' He went on to argue that the apartheid regime did society a service by building dams, since it maximised supplies in the good years 'so that we had something to rely on in the bad years'. He went on to add: 'I know one should not build dams nowadays' (interview, 11 November 2006).

The retention of 'existing lawful use' has prompted several academics and water professionals to voice their concern that the conversion of existing use rights into licences, which is the ultimate aim of the current legislation, will effectively entrench existing power relationships (see for example Schreiner & Van Koppen 2002). They argue that as the allocation of rights was based on riparian land holdings, which were almost exclusively white, the current 'lawful uses' are overwhelmingly in the hands of the white sections of the population, particularly white commercial farmers. Since the riparian doctrine did not apply to the homelands, where land holdings were primarily governed under systems of communal tenure, the notion of 'existing lawful use' does not, strictly speaking, hold in such circumstances though it does apply to 'customary' practices. The only other lawful way in which people living in communal areas could access water of quantities above the given threshold for productive use was through projects established by the homeland governments, such as irrigation schemes.

Although made out to be almost self-explanatory by some DWAF staff, by drawing on discourses of economic disruption if 'existing lawful use' was not retained, the idea was not without its critics. One of the senior members of staff in the Water Research Commission noted that, 'when I heard that they [the drafters] were going to protect the rights of existing users, it made me almost sick' (interview, 20 April 2006).

Licences and trading

Key to the whole Act and subsequent allocation reform was the idea of licensing. If not covered in any of the other categories, a potential user must apply for a licence. Unlike use rights under the previous legislation that could be exercised in perpetuity, licences are of a temporal nature: the maximum length that a licence can be issued for is 40 years, and it is subject to review every five years. A licence can be granted for any of the 11 uses of water defined in the Act, which also contains a 'disclaimer' in that it states that the issue of a licence to use water does not include a guarantee relating to either the statistical probability of supply, the availability of water or the quality of the water (see section 31 of the NWA; also Thompson 2006: 383). As noted, licences were regarded by some as conducive to institutionalising water markets. In particular, Gerhard Backeberg of the Water Research Commission is an ardent advocate of this (Backeberg 1995, 2005).

Originally, provisions for transferability, or trade, of water-use rights were only introduced in the third draft of the Bill and, according to one member of the drafting team, the government was dragging its feet over the issue. However, at the time of research, some members of staff within the department were beginning to embrace the idea of markets as vehicles for redistribution. In an interview with the Director: Policy and Strategy Co-ordination, he argued that 'it's not a matter of whether to introduce water markets, but of when and how...[we] shouldn't dilly dally; we need to do something soon'. The gist of his argument was that one could charge all water users in a region or basin. Bigger users would pay more, and the levy would be ploughed back into a 'reallocation fund', whereby the department could buy up water rights (interview, 20 April 2006). However, others within the DWAF saw this as far too expensive an option.

A key difference between the idea of permits in the 1956 Water Act and the licences of the 1998 Act is that permits were largely attenuated riparian rights, where the rights accrued from being in possession of riparian land. This attenuation generally conditioned the ways in which water was used but did not infringe on the right of use itself, though permits could also be issued to non-riparian holders to allow for industrial expansion. Licences issued under the 1998 Act are to a far greater extent subject to the discretionary power of the government, and are of a temporary nature, that is, their renewal is at the discretion of the authorities.

Whereas trade of permits or quotas was prohibited under the 1956 Act, such trades, or transfers, were now lawful and could take two forms. Temporary transfers of rights could take place without having to apply to the DWAF, as long as these transfers occurred within the same irrigation district and were approved by the Irrigation Board. Permanent transfers, where an existing user wished to convert the water use from a given resource (say from agriculture to industry), required an application for a permanent transfer, in which case the applicant would need to surrender the existing right and apply for a licence. A trade could be effected if another person applied for a licence to take up the surrendered right (see Appendix C), which would need to be authorised by the DWAF. In stressed catchments, where no new licences were issued and where the only uses recognised were existing lawful use and Schedule I, this was the only way anybody could get a licence for water use. In the Inkomati region, for instance, which was designated a stressed catchment, only one to two per cent of users had a licence (Deputy Director: DWAF Regional Office, interview, 8 September 2006). At that time, the DWAF received about 250 applications for licences annually, of which some 150 were issued. Most of these were applications for trading[23] or converting water use. According to the Director: Water Utilisation, each licence application was judged according to the criteria set out in article 27 of the NWA, and took on average 12 months to handle as a major constraining issue was the lack of information about Reserve requirements (interview, 15 August 2006). The length of application processing was also a recurring issue in the public hearings connected to the review of the NWA in October 2008 (Parliamentary Monitoring Group, 8 October 2008).

Compulsory licensing

In the NWA, licences were conceived in terms of facilitating reallocation of water resources through the process of compulsory licensing (sections 43-48). Basically, compulsory licensing is a mechanism whereby all the water uses in a specific area are cancelled and a call for licences is issued. This is primarily used in areas in which there is considered to be water stress, which was judged to be the case for 11 out of the 19 WMAs (Director: Water Utilisation, interview, 15 August 2006). According to the NWA section 43(1), undertaking compulsory licensing should be considered in order

    (a) to achieve a fair allocation of water from a water resource in accordance with section 45

(i) when the area is under water stress, or
(ii) when it is necessary to review prevailing water use to achieve equity of allocation;
(b) to promote beneficial use of water in the public interest;
(c) to facilitate efficient management of the water resource; or
(d) to protect water resource quality.

Once all the applications have been received, an allocation schedule must be drawn up and published in the *Government Gazette*. The schedule must outline in detail how water will be allocated among the applicants, with special consideration given to certain categories of these. Affected parties are provided with the opportunity of lodging an appeal. The responsible authority is under no obligation to allocate all available water, and may keep some water for future needs. Alternatively, it may allocate any surplus water remaining after the requirements of the Reserve, international obligations and corrective action have been addressed on the basis of public auction or tender.

Contested compensation: The 'safeguard clause'

Unlike land reform, where compensation is payable, the NWA contains a clause, 22(7)(b), that Van Koppen et al. (2002) have dubbed the 'safeguard clause'. This clause states that compensation may not be payable if water is taken away to provide for the Reserve, rectify an over-allocation, or rectify unfair or disproportionate water use.[24] Section 4(4) of the NWA states that the replacement of the exclusive use rights under previous laws does not require compensation to be paid. Van Der Schyff (2003) argues that the conversion of existing lawful uses into licences does not represent an expropriation of property, but rather a deprivation in the public interest. 'Expropriation' occurs when the State takes away a right in property and either keeps it for itself or transfers it to someone else. Deprivation implies that the right to property remains in the hands of the original holder, but that certain conditions or restrictions are imposed on the exercise of the right by the State.

It is important that deprivation should not be arbitrary. Something is arbitrary according to South African administrative law if it is 'capricious or proceeding merely from the will and not based on reason or principle' (Thompson 2006: 180).[25] Thompson seems to be in line with Van Der Schyff in terms of viewing the right of the State to infringe on the property of individuals in order to promote 'beneficial use in the public interest' (2006: 389–390), as long as the deprivation is not arbitrary. However, he goes on to argue that if 'entitlements are transferred due to the implementation of compulsory licensing procedures, it is submitted that it is an expropriation that should comply with the provisions of the Constitution' (2006: 390). Although the replacement of uses recognised under previous legislation with the new entitlements in the NWA constitutes a deprivation rather than an expropriation, Thompson regards the process of compulsory licensing as constituting an expropriation, as it involves – in his view – the taking away of water from some users to give to others.

Thus, the Act laid the foundations for allocation reform, in terms of defining categories – Schedule I, general authorisations, existing lawful use and licences – and allocation procedures. This seemingly straightforward exercise, however, was fraught with ambiguities and tensions, some of which have been hinted at above. These tensions provided the substance – the discursive spaces – for the emergence of particular narratives with respect to how rights should be allocated, and what principles and procedures should guide decisions. It was the fleshing out of these principles and procedures that constituted what became known as the Water Allocation Reform.

Notes

1. No. 27 of 1913, commonly referred to as the Land Act.
2. *Mfacane* (alternative spelling *Mfecane*) means 'the crushing' in isiZulu, whereas *Difaqane* means 'forced migration' in Sotho (Morris 1998).
3. The Cape is now part of the Western Cape, although by 1814, Dutch and British settlers had settled in what is now the Eastern Cape; Natal is KwaZulu-Natal in present-day South Africa; the Orange Free State is the Free State; and Transvaal was subdivided into four provinces in 1994: Limpopo, North West, Gauteng and Mpumalanga.
4. Boer means farmer in Dutch, and was the name under which the new colonisers became known. This term is used interchangeably with Afrikaner in this book.
5. The Pass Laws required Africans to carry an identity document in the form of a reference book, which contained details of their employment history and rights of residence.
6. The terms Bantustan and homeland are used interchangeably in this book. The apartheid government created Bantustans with the Bantu Authorities Act (No. 68 of 1951). The concept of homelands came into being with the Promotion of Bantu Self-Government Act (No. 46 of 1959), and the Bantu Homelands Citizenship Act of 1970. So the terms homelands and Bantustans are commonly interchangeably used, with political connotations, but we must be careful that we do not use the term homelands before 1951, or the term reserves after 1951.
7. Since many of South Africa's rivers are non-perennial, they could, under Roman-Dutch law, be subject to private ownership.
8. This idea is known in Latin as *Cuius est solum, eius est usque ad coelum et ad inferos*.
9. Roman *praetors* were annually elected magistrates who were responsible for enforcing the law enacted by the people. They set annual edicts that publicly set forth how the law would be enforced, both substantively and practically.

10 Roman law had almost been forgotten until its revival in Italy in the twelfth century, and over time was partially incorporated into the primitive Germanic laws of Western European countries, including the Netherlands. It was this 'hybrid bloom' that was brought to the Cape in 1652, and that came to be known as the Roman-Dutch law (Thompson H 2006).

11 The term Washington Consensus most commonly refers to an orientation towards neoliberal policies that was influential, from about 1980-2008, among mainstream economists, politicians, journalists and global institutions like the International Monetary Fund and World Bank. The term can refer to market-friendly policies that were generally advised and implemented both for advanced and emerging economies. It is sometimes used in a narrower sense to refer to economic reforms that were prescribed just for developing countries, which included advice to reduce government deficits, to deregulate international trade and cross-border investment, and to pursue export-led growth (from Wikipedia).

12 Restitution is primarily about restoring land rights to those dispossessed during the homeland consolidation processes and the attendant forced removals under the apartheid era. However, according to Hall (2004), only a handful of the more than 36 000 claims had been settled by late 2003, due to the complexity and costliness of settling claims in rural areas involving large numbers of people on large tracts of land.

13 In terms of redistribution, a willing seller, willing buyer principle was adopted, not so much because of World Bank influence, but because of domestic policy negotiations (Lahiff 2005). The policy has met with much criticism (see, for example, Lahiff 2005; Hall 2004; Cousins & Hall 2010) and has failed to meet the initial goals of land redistribution. According to Hall (2004: 214), the land reform evinced a 'spectacular underperformance'. The ANC committed itself, as part of the RDP approach to land reform, to redistribute 30 per cent of agricultural land to the poor and landless over a period of five years (Hall 2004). World Bank advisers had proposed this target as feasible, noting that six per cent of agricultural land is transacted each year, thus appearing to hold to the incredible notion that all, or nearly all, land on the market would be bought for redistribution (Aliber & Mokoena, 2002: 10). To provide a sense of scale, the commercial farming areas of South Africa amount to about 86 million hectares, and the land reform target was to transfer 26 million hectares in the first five years (Hall 2004). In the first five years of ANC government – what became known as the Mandela era (1994–1999) – only a small fraction of the target of 30 per cent was met. Subsequently, during the Mbeki era, there was a shift in focus from the rural poor to aspiring black commercial farmers.

14 The Legal Resources Centre and attorneys Webber, Wentzel & Bowens brought a legal challenge against the Act, arguing that it was unconstitutional. The Constitutional Court did eventually declare the Act unconstitutional on 11 May 2010. See also Claassens & Cousins (2008a) for further information on the process, which opened up a whole new rethink on South African land reform (see Cousins & Hall 2010).

15 This second phase included seven main criteria upon which the empowerment credentials of businesses in South Africa are assessed: ownership, management representation, employment equity, skills development, preferential procurement, enterprise development and corporate social investment (Ponte et al. 2007).

16 Asmal was succeeded by Ronnie Kasrils, with Buyelwa Sonjica taking over the office in May 2004. Following the death of Stella Sigcau, then Minister of Public Works, a Cabinet

reshuffle took place in May 2006, instituting the former Minister of Energy and Minerals, Lindiwe Hendricks, as Minister of Water Affairs and Forestry, thus effectively swapping portfolios with Sonjica. Many were critical of Mbeki's reasons for the Cabinet reshuffle, which saw the Minister of Agriculture and Land Affairs, Thoko Didiza, moving to take up Sigcau's post (*Mail & Guardian*, 23 May 2006).

17 The Free Basic Service policy stated that households should have access to 6 000 litres of water per month free of charge.

18 However, De Coning's use of the terms 'private' and 'public' goods with respect to water are here slightly confusing, given the 'closed commons' nature of the riparian doctrine system.

19 In the DWAF's presentation to the Parliamentary Monitoring Group in connection with the public hearings on the review of the National Water Act (8 October 2008), it was informed that two CMAs had been fully established while six had been enacted.

20 It is worth remarking in this context on the somewhat curious situation of the Department representing a lumped-together portfolio of 'water' and 'forestry', especially considering the vested interests of the forestry sector with respect to water use. Forestry should arguably be placed under the Department of Agriculture (DoA), as it is in most cases an industrially cultivated mono-crop (Andre van Tonder, interview, 28 November 2006; Philip Owen, interview, 25 September 2006). In May 2009, however, following the inauguration of President Zuma, the Department was renamed the Department of Water, and forestry issues were relocated to the Department of Agriculture, Forestry and Fisheries.

21 This is pointed out by Barbara van Koppen and Barbara Tapela, personal communication, (September 2006).

22 Section 6(1) of the Act reads as follows: 'Subject to subsection (2), no person may use water services from a source other than a water services provider nominated by the water services authority having jurisdiction in the area in question, without the approval of that water services authority.'

23 The two parties – the individual surrendering the right and the person applying to take up the surrendered right – negotiate the price themselves, but then the arbiter, that is, the DWAF, can state conditions that did not apply to the original owner. Thus, in a sense, the traders do not know the type of deal they are agreeing to before it has been approved by the DWAF.

24 Section 22(6): Any person who has applied for a licence in terms of section 43 in respect of an existing lawful water use as contemplated in section 32, and whose application has been refused or who has been granted a licence for a lesser use than the existing lawful water use, resulting in severe prejudice to the economic viability of an undertaking in respect of which the water was beneficially used, may, subject to subsections (7) and (8), claim compensation for any financial loss suffered in consequence.
Section (7): The amount of any compensation payable must be determined
   (a) in accordance with section 25(3) of the Constitution; and
   (b) by disregarding any reduction in the existing lawful water use made in order to
      (i) provide for the Reserve;
      (ii) rectify an overallocation of water use from the resource in question; or
      (iii) rectify an unfair or disproportionate water use.

25 Thompson is a lawyer, and also a member of the Expert Panel.

3

WAR in the making: Crafting the Water Allocation Reform Programme

THIS CHAPTER TRACES THE process of drafting the Water Allocation Reform (WAR)programme. It describes how the whole reform process came to centre on the idea of dealing with scarcity, and how the premise of managing a scarce resource holistically, created the need to register all water users. One of the main pillars of redistribution was the idea of compulsory licensing, whereby all existing and potential users would have to apply for a water use licence, and in the process priority would be given to those users who were historically disadvantaged. This chapter describes the setting up of the Expert Panel in 2003, and the early drafts of the WAR programme through to the final WAR position paper published in November 2006. The emerging perspectives on allocation issues and how these changed over time are mapped, showing how the use of certain terms – such as existing lawful uses – contributed to framing allocation issues in particular ways. The failure to link up with land reform processes is highlighted, and the chapter ends by contending that the insistence on maintaining existing lawful users effectively narrowed down the room for manoeuvre in reforming water rights distribution.

Water Allocation Reform: The basis and the process

As we saw in the previous chapter, Asmal's Act left the details of redistribution in the doldrums. The Water Law Principles (see Appendix B) were too vague to offer the necessary practical guidance in terms of water allocation, so there was a need to flesh out more precisely just how water would be redistributed. It was recognised that a separate, more pragmatically oriented policy was needed to spell this out. This was eventually to become the WAR position paper. In this chapter, I highlight how the idea of scarcity and the associated need for accurate quantification of the available water resources came to dominate the early stages of the reform process, and how water use registration and the notion of compulsory licensing came to form the basis for redistribution. I go on to describe briefly the actors and the processes involved in drafting the reform, and analyse the contents of the reform documents, highlighting how water users were categorised and associated with particular characteristics. I then tease out the implications of this.

Through the lens of scarcity

Scarcity was a major theme throughout the drafting of the Act, as the following passage in the White Paper demonstrates:

> South Africa is an arid country with rainfall less than the world average very unevenly distributed across the country. With just over 1 200 kℓ of available fresh water for each person each year at the present population of around 42 million, we are on the threshold of the internationally used definition of 'water stress'. Within a few years, population growth will take us below this level. South Africa already has less water per person than countries widely considered to be much drier, such as Namibia and Botswana.
> (White Paper on a National Water Policy for South Africa: 1997a: 13)

This passage, which uses the amount of fresh water per capita as a point of departure to declare that South Africa is a water-scarce country, draws on work by Falkenmark and others (Falkenmark & Widstrand 1992; Falkenmark 1998) that create indices for water stress and water poverty through aggregating a country's water resources and dividing this by the population. The British economist and demographer Thomas Malthus did the same thing with land and food production over 200 years ago. In his influential work, *An essay on the principle of population*[1] (published in 1789), he argued that population growth would always outstrip productive capacity, and hence facing scarcity would be the permanent plight of humankind. Cairncross (2003) takes issue with the fact that Falkenmark only took account of water that could be abstracted from river flows and neglected the vital importance of rain-fed agriculture in many arid and semi-arid areas. Moreover, his focus on flows, rather than stocks, ignores large lakes and water reservoirs. Finally, the use of the term 'used up' detracts from the fact that water is diverted, transformed and polluted, but that this permits recycling, cleaning and reuse. Hence, such accounts of physical scarcity are framed in a neo-Malthusian vein. They are rather simplistic and reduce the multi-dimensional phenomenon of water availability – involving social, political, meteorological, hydrological and agricultural factors – to an issue of physical quantification. Moreover, they fail to say anything about orders of scarcity, meaning that there is a distinction between what Ohlsson (1998; 1999, quoted in Turton & Ohlsson 1999: 2) has termed a 'first-order scarcity of natural resources' and a 'second-order scarcity of social resources'. This second-order scarcity may be perceived as a dearth of adaptive capacity.[2] This is particularly pertinent when considering the argument that the available freshwater resources of the Southern African Development Community (SADC) are enormous (Boroto 2004) – the problem is that there is such a high degree of variability in flows, with frequent floods and droughts, and many people live in densely settled areas far from where the water is most readily available (Turton, Schultz et al. 2006). These factors lead to situations of localised scarcity. The water scarcity discourse in South Africa, however, downplays such aspects and regional differences. Although there is undeniably a physical scarcity in many of South Africa's catchment areas, the idea of scarcity is 'nationalised' through the use of country-wide averages.[3] Scarcity becomes not only nationalised, but is also naturalised (Mehta 2005) in the sense that it is taken as a given, as a fact of life and as an environmental orthodoxy (Forsyth 2003). Hence, the strong focus on physical quantification downplays the multidimensional nature of water management.

As observed in the introductory chapter, framing serves to organise thoughts on a certain issue by highlighting some factors and downplaying others, thus demarcating what counts as relevant in a particular context (Rein & Schön 1993). Framing the issue of water access mainly in terms of a problem of physical dearth works to preclude other perspectives. Whereas the drafters of the NWA had been preoccupied with quantity concerns, quality was relegated to the back seat in the process of drafting guidelines for reform. The issue of quality was only included in the final draft of the WAR guidelines (NWRS author, personal communication, 17 May 2006; Director: Water Allocations, interview, 31 October 2006). Hence, the emphasis came to rest on the need to quantify the available water in order to facilitate reallocation of use rights. The focus on compulsory licensing as a means of redress was the key, and it dominated much of the reform process. A premise of compulsory licensing, however, was the assumption that it could not proceed without an accurate assessment of available water resources.

The pillars of reform: Registration and compulsory licensing

The Water Authorisation and Resource Management System
    The first step towards implementing the NWA and facilitating WAR was to register all existing users. The registration of water users commenced in 2000,[4] when the then Minister of Water Affairs and Forestry, Ronnie Kasrils, delegated the power to register water use and issue registration certificates to the Regional Directors. All water users over and above Schedule I were required to register their use. For the purpose of handling the registrations, DWAF created a database – the Water Authorisation and Registration Management System (WARMS). Registered

users were issued with a Registration Certificate, but this certificate provided no entitlement as such: it merely acknowledged the fact that a particular individual was currently using a specified amount of water. The ultimate aim of WAR was to convert all certificates into licences. However, as Garduño and Hinsch note, there were several problems with WARMS. Many people apparently thought that the best strategy was to 'lay [one's] hands on water now, otherwise the Reserve will take it all' (Garduño & Hinsch 2005: 53), and thus over-registered their use. This was particularly evident in the Inkomati WMA, where the estimated use in the NWRS amounted to 844 million m³ per year (see Chapter 4), whereas the use that was registered totalled 1.4 billion m³ per year.[5]

The WARMS database for the Mpumalanga region, covering the Olifants and Inkomati WMAs, was organised by districts rather than catchment boundaries, rendering it difficult to sort data by WMA, and underscoring the point made by Chereni (2007), among others, that it is difficult to align the 'institutional principle' of catchment management with existing institutional structures of governance. The database basically contains information about the source of water (the Komati River, Lomati River, and so on) and the purpose and quantity of water registered by the users. It contains no explicit information about race or gender and only categorises users (called 'customers') as an individual, a company, a water user association or a service provider. In the case of agriculture, information was provided on how many hectares are irrigated, the type of crops for which water is registered, the start date for planting and growing days per season, and the type of irrigation system used.

The figures in the database were very precise, for example, water volumes were given in the format '9 833 m³'. In addition, interestingly, water source entries such as 'natural rainfall' (for a 30-hectare plot of oats), and even one-hectare plots (vegetables), were registered. Starting in 1980, plots of five hectares of mealies and vegetables were all listed as using 80 400 m³ per year from the Komati River, and other uses (from boreholes) were listed as drawing 26 000 to 31 000 m³ of water per year. Although Schedule I users did not need to register, there was an entry for a Schedule I use from a borehole in the Crocodile WMA from 1901, using 183 m³ per year. More Schedule I users were registered as taking up water use in 1996, using between 2 000 and 5 000 m³ a year, which are minor quantities. Even 0.5-hectare plots of vegetables, allegedly using 26 285 m³ of water per year, were registered.

Though the ultimate aim of WARMS was to create a comprehensive database of licensed water users, the initial purpose was primarily billing. The registration of water uses followed the Pricing Strategy, published in 1999 (DWAF 1999), which outlined the cost of water use differentiated by geographical location and sector. WARMS was a means to charge users accordingly. A draft version of an internal DWAF document on 'Economic considerations of water allocation' argues on page 17 that, 'the raw water pricing strategy is designed to recover the costs of management of the water resource and water resource development and is not seen as a scarcity charge to allocate the resource'. It has little to do, then, with the 'demand management' approaches of the reform process.

Compulsory licensing

> The correct assessment of the availability of water (both surface and groundwater) for allocation to the various users in each affected catchment is the cornerstone of the compulsory licensing process.
> (Ninham Shand, accessed 19 June 2006, http://www.shands.co.za)

The WAR was intimately bound up with the idea of compulsory licensing. In essence, this entailed cancelling all existing water-use rights in a 'stressed' catchment area and issuing a new call for licences. All existing and potential users were subsequently invited to submit applications for water use licences. These would then be issued based on certain criteria that included applying the ideals of affirmative action, thus emphasising the importance of issuing licences to historically disadvantaged individuals (HDIs). Compulsory licensing was one of the pillars of the allocation process.

In the quotation from the Ninham Shand website above, scientific accuracy is portrayed as the linchpin of compulsory licensing. The quote conveys an implicit assumption of hydrologists' ability to produce accurate answers, which belies the internal contestations, ambiguities and uncertainties besetting scientific approaches to estimating water availability, abstraction and allocation.[6] The use of the term 'correct' reflects what Potter observes, namely that 'scientific discourse has the particular characteristic of describing the world in an 'out there' fashion, as objective truth, without agency, subjectivity and uncertainty' (1996, quoted in Wester & Warner 2002: 67). Uncertainty and contestations were rife in terms of modelling approaches and the optimal way of going about the estimation of water flows and abstractions. In the Department for International Development (DFID) consultant's opinion, DWAF was divided into factions when it came to modelling approaches, with the old guard of modellers acting as gatekeepers and holding the allocation process hostage in their desire to get the models right first by playing around with different scenarios. 'The problem with this is that everything then has to go through this gate, and it's going to take years and years to generate the information required' (interview, 18 May 2006).

Hence, the emphasis on scarcity and holistic management generated a belief that a sophisticated system of quantification and registration was needed to gain an overview of the available water resources and their use, and that compulsory licensing could only be carried out once water rights had been accurately identified.

Processes, actors and perceptions

Who were the institutions and people driving the process? And how was it carried out? The account that follows is by no means exhaustive, but aims to give an impression of some of the key players, in order to understand how these shaped the emerging perspectives and discourses around water allocation. Thus, before going on to analyse the discourses themselves, it is important to understand the 'clever, creative human beings' (Hajer 1995: 58) involved in shaping those discourses.

The Department of Water Affairs and Forestry and the Department for International Development

DWAF had had both its coffers and its staff levels boosted after 1994 and grew to occupy a strong position, both nationally and internationally. Its emphasis on equity and redress easily attracted donors' attention, such as DFID. The process of drafting a strategy for WAR was part of the DFID-supported Water and Forestry Support Programme (WFSP). DFID had been supporting South Africa's water sector since 1995, and the WFSP was a means to further extend that support. It was developed in collaboration with DWAF and totalled some £19.8 million. The WFSP had five components: the water and sanitation services sector; water resources; corporate institutional transformation; strengthening the forestry directorate within the DWAF, and a focus on 'making forestry markets work for the poor'. The focus of the research informing this book falls largely within the scope of the second component of the programme, water resource management (WRM), which was funded with £4 million over a period of five years. The WRM component had two main aims: firstly to develop methodologies for water use allocation and monitoring that involved the rural poor, and secondly to test these methodologies in pilot catchments. The component had as its core the objective of fleshing out the principles of allocation that were inherent in the Act, and of making these realisable. Within DWAF, it was the Director: Water Allocation who became responsible for driving the whole process of fleshing out such methodologies for allocation. The consultant appointed by DFID to oversee the WRM component played a key role in this regard, and worked in tandem with the Director to move the process forward. The two of them were the signatories of the Project Implementation Unit (PIU), which was responsible for approving procurement of consultancy services and overseeing the whole process.

One of the five outputs – output 3 – of the WRM component was to produce a toolkit of allocation methodologies and monitoring procedures which would take particular cognisance of the needs of the rural poor. Output 3 involved the development of generic allocation procedures, which were intended to be completed by the first quarter of 2004. Developing generic allocation procedures was interpreted as describing how the Act would be implemented in practical terms, which included setting up an expert advisory panel that would act as a sounding board in the formulation of allocation principles. Advice provided by the board would serve as a 'key input in terms of vetting the products emanating from the various projects' (minutes, first Expert Panel meeting, 20 November 2003).

The Expert Panel and other actors

The Expert Panel met four times during 2003 and 2004, which was in the early days of the process of drafting the position paper. Attendance varied at the four meetings. Provided below is an overview of people who were present at all, or at least three, of the meetings, to give a feel for the composition of the panel.

Only eight people attended a minimum of three of the four meetings:
- the Director: Water Allocation from DWAF (male);
- the DFID consultant responsible for the WRM component (male)
- three representatives from different consulting firms (all whites, two male, one female);
- a black, female representative from the Water for Food Movement (WFM);
- a white, female, senior researcher from the International Water Management Institute (IWMI); and
- a white, male, senior researcher from the Council for Scientific and Industrial Research (CSIR).

Nine people attended a minimum of two meetings:
- two lawyers (both white males);
- a white, male senior researcher from the Water Research Commission (WRC);
- a female, black member of parliament; and
- several consultants.

A representative from the Programme for Land and Agrarian Studies (PLAAS) at the University of the Western Cape was also invited, but was unable to attend because of personal commitments. No substitute was found, and so the panel did not have any representatives who were knowledgeable about the land reform process.

Initially the debate, as reflected in the minutes of the meetings, centred on what the task at hand actually was. Was the idea to develop a strategy, or a framework? Was it a policy or not? An initial suggestion put forward was that of viewing the task as producing a 'covenant' between the State and its 'customers', but eventually the group settled on the less austere-sounding term 'position paper', which was later to be dubbed 'guidelines', and then, with the publication of the final version in November 2006, it had turned into a 'strategy'. This was a far more active-sounding appellation than the initial term 'position paper', and reflected the increasing urgency on the DWAF's part with respect to the progress of reform.

Both the lawyers on the Expert Panel had been part of the drafting process of the NWA in various capacities. The consultants were mostly environmental specialists. One of them had been intimately involved with the drafting of the NWA and took on the responsibility of developing and implementing regulatory measures in terms of protecting aquatic ecosystems, but later moved abroad and was no longer involved.

The CSIR representative hailed from an institution at which water-related research tended to have an environmental bent. Its research activities with respect to water were located within other research themes such as ecosystems management, environmental impact assessments, and the linkages between water, land and marine ecosystems. Its focus was on 'the optimal utilisation of natural resources in support of economic growth and human well-being'.[7] The CSIR representative was also linked to the African Water Issues Research Unit (AWIRU), a not-for-profit organisation associated with the University of Pretoria, which focused on 'generating water management solutions that were politically, socially, economically, environmentally and culturally sustainable'.[8]

Another important representative on the panel was the Water Research Commission (WRC). The Commission was established in 1971, in accordance with the Water Research Act.[9] It played a key role in knowledge production, and served as the national knowledge hub for water. Its mandate was to support water research and development, as well as the building of a sustainable water research capacity in South Africa. This included the creation, dissemination and application of water-centred knowledge, which focused on water resource management, water-linked ecosystems, water use and waste management, and water utilisation in agriculture. The Director of the Water Utilisation in Agriculture section of the WRC was an ardent advocate of water markets, which was also reflected in its research portfolio.

The IWMI, on the other hand, was a newcomer. It established a regional office (the head office being in Colombo, Sri Lanka) in Pretoria in 2000, two years after the Act had been passed. Its mission is to improve 'water and land resources management for food, livelihoods and nature'.[10] At the time, many of IWMI's research efforts focused on the Olifants River Basin, as this was a benchmark basin in the Consultative Group on International Agricultural Research (CGIAR) Challenge programme, in which IWMI was heavily involved. It emphasised collaborative research and partnerships and therefore liaised with government departments, NGOs, research organisations and networks. At the time, the IWMI South Africa office was liaising more with the then Department of Agriculture

(DoA – now the Department of Agriculture, Forestry and Fisheries) than with DWAF, reflecting its agricultural focus.

Closely associated with IWMI was the Water for Food Movement (WFM), a small NGO concerned with strengthening food security through implementing rainwater harvesting programmes in rural villages. Another important NGO working in the rural areas (mainly in Limpopo and Mpumalanga) was the Association for Water and Rural Development (AWARD), which focused on water and livelihoods. Although they had no formal involvement in the policy process, they were part of the informal network and were also contracted to work on particular DWAF assignments.

Even though IWMI played no formal role in policy-making, its senior researcher and leading social scientist was much sought after nationally and internationally and was invited to sit in at the formal meetings at DWAF (IWMI SA office Director, interview, 6 June 2006). She had earned herself a reputation among other researchers as 'something of an activist' (CSIR/AWIRU researcher, interview, 2 May 2006; University of Pretoria researchers, interview, 14 April 2006). For her part, she perceived several academics associated with the University of Pretoria and other institutions as being only interested in models (personal communication).

Apart from inviting them to take part in formal meetings, DFID/DWAF also commissioned IWMI, along with the consultancy Ninham Shand, as one of the contributors to the development of tools for monitoring the progress of allocation reform, which gave rise to the innovative idea of using the concept of the Gini coefficient to monitor levels of water inequality in a basin (see Cullis & Van Koppen 2007).

IWMI's position thus provided a counterweight to CSIR/AWIRU-led research, and there was little attempt to hide the fact that these institutions viewed each other as rivals in terms of influencing policy. IWMI was clearly the outsider, and some CSIR/AWIRU staff resented the fact that IWMI had come to play an influential role in the forging of government policy.

Consultation and participation

In spite of DWAF's emphasis on consultation, there were critical voices raised with respect to these processes. One (black) external consultant working on the Water Resources component of the programme argued that the panel was largely a 'consensus-seeking' forum which was not amenable to revealing the 'real' problems or controversies. These, he maintained, would be better exposed if DWAF had arranged to meet with individuals to have one-on-one talks in combination with smaller, more informal groups. The CSIR representative called the bringing together of an Expert Panel merely a 'rubber-stamping' process that was carried out to 'keep the donors happy'. In his view, it was not really a consultative process as there was no space for learning. The fact that other processes were going on in parallel was hampering the ability of the experts to bridge gaps (email communication, 17 November 2007).

There were misgivings within the Department on the style of the drafting process. A key staff member who explicitly asked to remain anonymous ('in order not to ruffle feathers') was unhappy that the policy wasn't discussed across the board in-house before being taken on a round of consultations, and in this regard said:

> It is, for me, a serious problem that DWAF staff, Allocations and Planning, for example, is not meeting to discuss approaches across the table, and that 'policy' of this nature is being developed in isolation and finally presented in an indaba.[11] It must be said that an earlier version of the WAR [Water Allocation Reform] document was taken around to stakeholders in all the regions early last year, but there again I was very bothered that it had not yet been discussed across the table within DWAF. The consequence was that, in my opinion, a very flawed document was taken out to stakeholders, and that many of the problems could have first been sorted out in-house.
> (Interview, 26 April 2006)

This impression was reinforced by the DFID consultant's own description of the drafting process as something that was 'locked up in the heads of [the Director: Water Allocation] and me...we have had long discussions on nuances' (interview, 18 May 2006). Knowledge making was thus internalised within a group of a few individuals. Although IWMI, CSIR and others were officially taken on board through their participation on the Expert Panel, there was little doubt that it was DFID and, ultimately, DWAF who called the shots. The indaba referred to above was convened in March 2006 to discuss progress on the WAR. Even the DFID consultant was not allowed to participate in this, as he was external to DWAF.

The position paper was made public in January 2005. The process of taking the flawed document out to stakeholders (see above) occurred over a nine-month period in 2005. Nine provincial stakeholder meetings were convened, with the consultation process culminating in a national consultative meeting that was held in Pretoria in April 2005.

While these meetings generated comments from different sections of society, they reflected a strategy of two-way communication rather than real participation, as stakeholders' views were sought on already defined issues. Stakeholders did not actively participate in the identification of issues themselves as the frame and focus of the main issues were already in place, and what they were expected to do was to comment on its content.

Emerging perspectives

The 'industrialist/institutionalist' perspective

The CSIR representative regarded the relatively recently established IWMI as something of an upstart, and commented that its staff did not know the history and context of South Africa: 'they only focus on irrigation' (interview, 2 May 2006). In this way, IWMI's legitimacy was discredited through invoking their perceived lack of sensitivity to local issues and historical knowledge, and their focus on water for irrigation only. In the CSIR representative's view, himself an Afrikaner, the Boers had been driven by a desire to make the desert bloom, to cultivate what they regarded as their God-given land, emulating the Zionist zeal of Israel. Water projects were seen as the means to perform miracles. Through massive resource capture, with the help of the government's 'hydraulic mission' (Swyngedouw 1999, in Turton & Meissner 2002: 41), these dreams were realised in the hinterland. Now, he continued, these romantic images are being re-invoked, but this time for emerging farmers. In his opinion, these would fail, mainly due to lack of capital, markets and financial resources. He countered what he claimed were 'romantic ideas about agriculture' and 'rural nonsense' by stating that 'South Africa is not an agricultural country...we need to transform ourselves from being a nation dependent on mining and agriculture, to one relying on industry'. Although he agreed on the necessity for redress, he argued that 'ultimately, we need to attract foreign direct investment' (FDI). In his opinion, foreign investors perceived an increasing level of uncertainty and risk as a result of the lack of clarity of the WAR. He cited the case of a client of the CSIR who had been unable to get a water licence, so had withdrawn a R3.5 billion contract. According to him, the reform process was sending potential business partners scuttling and was creating low investor confidence. The primary aim of the reform should have been to make sure that licences were made secure and reliable, in order to promote investor confidence and boost the South African economy (interview, 2 May 2006). This view speaks directly to the GEAR macroeconomic strategy of boosting the economy through attracting foreign capital (see Chapter 2, page 24).

In the words of Rein and Schön, 'Policy frames and their underlying appreciative systems are revealed through the stories participants are disposed to tell about policy situations' (1977, quoted in Rein & Schön 1993: 148). They go on to coin the term 'problem-setting stories'. In this instance, the representative from the CSIR was framing the problem in terms of how to facilitate the transition from agriculture to industrialisation, and how to make licences secure in order to attract FDI. He thus represented ideas from what may be dubbed an 'industrialist/institutionalist' discourse. It reflects a set of ideas and concepts that revolve around the notion of transforming South African society into being less dependent on primary sectors, and of seeing water use licences primarily as providing holders with security, which in turn will increase the likelihood of their making further investments in the country. This view is 'institutionalist' in the sense that it emphasises the high transaction costs incurred by high levels of uncertainty, which can be countered through the security afforded by the institutionalisation of water use licences.

The 'industrialist/institutionalist' discourse, therefore, sees licences in terms of their potential to offer security and promote water trade in favour of the industrial sector, thus conceiving the State's role primarily as being a guarantor of security. By construing redistributive efforts in terms of their potential to disrupt existing economic interests, this discourse largely favoured existing users, but underscored the potential of trade as an incentive to transfer agricultural water towards industrial uses.

The 'agriculturalist/livelihoods' perspective

In sharp contrast to the above perspective is the pro-agriculturalist stance, which was most expressly advocated by IWMI. Its livelihoods focus gradually grew more prominent and adopted emerging new ideas, such as the multiple-use systems (MUS) approach. This approach highlights the integrated nature of water in rural households: water from multiple sources is used for multiple purposes, including domestic and small-scale production uses (Moriarty, Butterworth et al. 2004). This approach focuses far more on the integrated aspect of the IWRM paradigm, but concentrates on the household level rather than the basin level.

The increasing focus on small-scale agriculture and livelihoods raised concerns with respect to the role of statutory law in general and formal licences in particular, and the overall role of water in sustaining livelihoods. IWMI helped to convene workshops on water use in agriculture and also on customary water rights and legal pluralism. A key event was the conference on water and legal pluralism with a focus on Africa (the African Water Laws conference), which was held in January 2005 and attended by a host of lawyers, managers, policy-makers, NGO representatives and academics. The conference was convened as part of a collaborative three-year research project that focused on legal pluralism and integrated water resource management (IWRM) through action research, capacity building and advocacy, a process in which IWMI was a key partner.

The papers presented at the conference dealt with cases from a number of African countries (Kenya, Tanzania, Zambia, Zimbabwe, Malawi, Ethiopia, Ghana and South Africa), but also included a paper from the Andes region. However, whereas several case studies stressed the strength of existing customary arrangements (such as the practices of the Shona in Zimbabwe and the Oromo people of Ethiopia), two of the four South African papers largely skirted the topic of customary rights and were more concerned with the details of establishing CMAs (Anderson 2005; Pegram & Bofilatos 2005). Another paper pointed to the problem of a mismatch in institutional boundaries and the complexity resulting from this (Pollard & Du Toit 2005). The only paper that purported to deal with 'traditional' water management was a case study of several villages in Limpopo, based on brief field visits and co-authored by a team of researchers from the African Water Issues Research Unit (AWIRU) and the CSIR (Malzbender, Goldin et al. 2005). Hence, there is a dearth of knowledge about, and an obvious need for, more detailed research of the nature of customary water use in South Africa, particularly because the apartheid era effectively hindered the establishment of a robust African peasantry.

Land tenure is a far larger issue than water rights, but scant attention was paid to the politics of land reform, which reflects how research related to the water reform and land reform processes were not 'talking to one another' at the time.[12] In the plenary statement, a brief reference was made to the contested authority of traditional chiefs, arguing that this would be countered through the processes of democratisation (Water Laws Workshop 2005), but none of the papers presented at the conference effectively engaged with the issue.

The conference produced some important recommendations, among them the need to raise the threshold of registration requirements so that those needing water for livelihood purposes did not have to go through the cumbersome and bureaucratic process of obtaining a licence. Inspired by multiple-use thinking, the need to better integrate domestic and small-scale productive uses was emphasised.

However, suffusing the debates on customary law were discourses on the ideals of collective action, and of communities as being inherently more equitable and capable of sustainable management. The plenary statement, for example, argued that 'water uses were and are governed by local community-based water arrangements, which are embedded in a community's holistic governance of human and natural resources' (Water Laws Workshop 2005). This was rather sweeping, and also striking in light of the paucity of empirical material with respect to documenting customary water laws in South Africa.[13] The livelihoods discourse, then, evinced an ingrained scepticism of the State and the statutory, positing the community as the locus of action, but in so doing it ran the risk of eulogising and collapsing into monochrome a concept that is full of nuances.

The livelihoods perception of licences regarded licences with trenchant scepticism, rather than seeing them as State-backed security. The strongest advocates for the livelihoods perspective saw licences as tools of deceit and dispossession, which served to entrench existing power patterns rather than produce new opportunities for the rural poor. This view was mainly founded on the belief that the administrative processes that had to be gone through in order to obtain a water-use right would favour those who were educated and well versed in the ways of administrative bureaucracy, and thus cause the poor and those less well educated to suffer a severe disadvantage. In addition, licences were envisaged as an administrative nightmare in terms of necessitating the registration of swathes of smallholders scattered throughout the rural areas. The issuing of State-backed licences, then, was regarded as a process of entrenching power in the hands of the already powerful.

Such discourses do not grow out of a vacuum, but are influenced by the political economy conditions within which they emerge. The 'institutionalist' discourse, in particular, was heavily influenced by the macroeconomic policy shift that had occurred shortly after the transition, which saw the radical RDP give way to the much more hard-nosed neoliberal strategy of GEAR (see Chapter 2, page 23). While this view was in favour of protecting the property rights of existing users and fending off State interventionist elements of the WAR, this also created a dilemma. Given the fact that the main thrust of the 'industrialist/institutionalist' view was towards speeding up South Africa's transition from an economy reliant in the main on primary industries such as agriculture and mining towards a more industrialised society ready to defend its place in the globalised world of commerce, this implied an uneasy relationship with the existing property rights holders, that is, the commercial farming sector. This was one of the most powerful interest groups in terms of water, accounting for some 60 per cent of water use, whilst agriculture's direct contribution to the GDP was estimated to be a mere 4.5 per cent[14] (DWAF 2004a: 30). Clearly, the path to industrialisation and growth meant getting water out of the commercial agricultural sector. How could the idea of protecting the rights of existing users – most of them commercial farmers – be reconciled with the urge to minimise their stake in water resources? One solution would be to encourage trade, and rely on the idea that farmers would willingly trade their water rights with industry if the price was right (this assumption, as Chapter 4 will show, did not always hold true.) Where the 'institutionalist' discourse took an overt neoliberal stance and was hand-in-glove with the GEAR policy, the livelihoods discourse took a bottom-up view and argued that development and reparations of past injustices was best achieved through providing communities access to water-use rights that could support diverse livelihood activities such as home gardening and small-scale agriculture. This view had much in common with the RDP in terms of its redistributive emphasis, but failed to deal adequately with the nature of the political power struggles that ensued after the transition and the associated negotiated settlement, in particular the matter of traditional authorities. It did not sufficiently locate its ideas of 'the community' in the current political context, which meant that it failed to appreciate – at least on the face of it – the power struggles that were taking place in terms of the ANC's attempts to accommodate traditional leaders in order to strengthen its rural base. Neither was there a serious attempt to deal more in-depth with localised dynamics, such as chiefs' authority vis-à-vis community members, or the very idea of 'community' itself. Given the nature of the forced uprooting and labour-migration systems (see Chapter 2, page 16) of the past, and the current migratory patterns which have given rise to bouts of xenophobia, the idealised idea of stable communities needs interrogation.

Whilst the DFID representative noted that 'both views are useful', there was dissension within DWAF itself in terms of where the emphasis should be placed when it came to the WAR. The above highlights how the concept of use rights, and in particular licences, were not givens, nor mere reflections of established relationships, but were actively formed through the ways in which particular terms and ideas were contextualised and given meaning. The next section examines how these different interpretations engaged in a struggle for dominance.

Water Allocation Reform: On paper

Law cannot be dissociated from its social and political context: one needs to understand whose ideas and perspectives are perpetuated and institutionalised. The constitutive approach to law sees it as a 'pervasive influence in structuring society [more] than as a variable measure whose impact can be measured. Law is seen as a way of organising the world into categories and concepts, which, while providing spaces and opportunities, also constrain behaviour and serve to legitimate authority' (Garth & Sarat 1998: 2, quoted in Oomen 2005: 26). But 'power emanates in many places' (Oomen 2005: 22), and it is in discourse that power and knowledge are joined together to create categories, subjects and divisions (Foucault 1980). Law and policy do not only mirror and create power structures, but also make for meanings and understandings of identity (Cheater 1999; Gohen 1992: quoted in Oomen 2005: 22). This is a key point: that law and policy help to fashion certain identities. In the sections that follow, I will go through the various revisions of the water reform policy documents in some detail, highlighting how water users are categorised and associated with certain characteristics through the use of particular terms and concepts.

From WARP to WAR

In the initial versions of the position paper, the front page declared that it was 'a Position Paper for the Water Allocation Reform Programme (WARP)'. Later versions dropped the 'Programme' and turned the appellation into the much more aggressive-sounding WAR. The rationale for this somewhat belligerent turn was, as the then Minister for Water Affairs and Forestry, Buyelwa Sonjica, put it during the Water Summit in 2005, 'to make WAR on Poverty'. The metaphor certainly struck a chord, for some a dissonant one, and many actors in the reform process were deeply unhappy with the name change. In the words of the CSIR representative on the Expert Panel:

> It [the allocation reform] is a world-class policy, but the acronym is going to fudge it all up. I am amazed at the short-term memory of this country. Not long ago blood was flowing down the streets, thousands of people were victims of violence and we were close to civil war, and then they start a reform by declaring war. It is insane...The WAR is not in the spirit of reconciliation. It's much more about confrontation...what the WAR is doing is invoking the old primordial instinct of the Afrikaner to fight for their rights, to fight with guns to protect their resources.
> (Interview, 2 May 2006).

This shift illustrates how language is used as 'symbolic capital' (Bourdieu & Thompson 1991) in the policy context. It draws attention to the explicit ways in which language embodies power. Moreover, it reflects the war raging amongst the people involved in the policy-making process. It is a war not over resources, but over discourses, deploying as ammunition concepts, terms and words to tell persuasive stories.

Policy narratives and the construction of social identities

Below is the first passage of the introduction to the final WAR paper – now called a strategy – in order to give the reader a sense of the style and tone of the final document:[15]

> As custodians of the national water resource, the Department of Water Affairs and Forestry must promote the beneficial use of water in the best interests of all South Africans. In order to do this, water allocations must be carried out in a manner that promotes equity, addresses poverty, supports economic growth and provides opportunities for job creation. The allocation process recognises that redressing the effects of previous discriminatory legislation is necessary for social stability and to promote economic growth. Moreover, the water allocation process must allow for the sustainable use of water resources and must promote the efficient and non-wasteful use of water...

The final position paper focuses on three main water uses out of the 11 described in the Act, namely water abstraction, water storage and stream flow reduction activities. It is divided into three sections. The first section is an introduction, describing the purpose of the paper and explaining key terms and concepts. The second part outlines the main principles, presented as eight guidelines, against which water allocation processes can be tested. The third part deals with approaches to water allocation, and identifies three types of situations, namely i) catchments where the allocable water is sufficient to meet the needs of the applicant, as well as demands from other users in the foreseeable future; ii) catchments where the application(s) may exceed the allocable water; and iii) catchments where the water resources are already over-allocated and have been prioritised for compulsory licensing.

A remarkable revision of the document was the complete removal of the section on monitoring and enforcement. Whereas the first few drafts contained a separate section dealing with this (up to the January 2005 version), by 2006 it was completely gone. The author of the National Water Resources Strategy (NWRS) was quite blunt about the gradual marginalisation of monitoring from the position paper:

> DWAF is not the department to deal with measuring poverty. We only make water available and believe that by doing so, we are making people better off...there's no way we're going to be able to quantify that specifically in socio-economic terms...we've spent hours debating indicators, monitoring... we haven't even got any proxies.
> (Interview, 17 May 2006).

The following section highlights how the narrative emerging in the policy texts worked to create certain subject 'positionings', that is, the text defines particular categories, such as 'existing lawful uses' and 'historically disadvantaged individuals'. The text also creates putative causal links involving these categories, such as through highlighting the risks involved in providing HDIs with water (see the following section).

Existing lawful uses and historically disadvantaged individuals

Language is not neutral. Language is not only a means of communication, but also 'an instrument of power' observes Fairclough (1995), drawing on Bourdieu and Thompson (1991). One of the ways in which this instrument of power works is through the 'subject positionings' fashioned by discourse. The idea of a subject position is explained quite well by Fraser (1989: 295), who argues that discourses include 'narrative conventions available for constructing the individual and the collective stories that are constitutive of people's social identities'. By constructing particular versions of the world, and by positioning subjects in particular ways, discourses limit what can be said and done.

Through the exercise of categorising water uses, users are created correspondingly, and thus subjects are being created in the process of transformation, and are being invented and reinvented. Where before there were only irrigators, the Act has now created a host of new categories and correspondent subject positionings: stream flow reducers, existing lawful users (ELUs), HDIs, potential water users, and so on. Whereas the term ELUs is a new creation of the legislation, the term 'historically disadvantaged individuals' (HDIs) came into use after the demise of the apartheid regime and reflected the democratic government's goal of addressing inequalities through affirmative action. It reflects a core contention of social sciences that identities are construed rather than given, which is elaborated through the idea of subject positionings.[16]

The WAR largely revolves around these main categories, that of ELUs and that of HDIs, with an additional category of the 'commercial HDI'. Policy texts are 'social spaces in which two fundamental social processes simultaneously occur: *cognition* and *representation* of the world and social interaction' as argued by Fairclough (1995: 6, emphasis added). The 'cognition' in this context involves the infusion of meaning through the emergence of particular storylines that create the effect of fleshing out an intuitive subject positioning. The notion of ELUs was formed through the process of creating a 'chain of equivalence', that is, by emphasising what is equal, what is held in common, by a diversity of actors (Laclau 1996), which causes quite disparate social identities to be collapsed into one big 'bag' or category, in order to serve a particular purpose. In Laclau's words, 'the more the chain expands, the more differential features of each of the links will have to be dropped in order to keep alive what the equivalential chain attempts to express' (Laclau 1996: 208). The equivalential chain, in this instance, is all the different water users: mining, agriculture, industry, domestic and urban. These are all collapsed into the basket term ELUs, thus converting possible subject positionings into the dual set of ELUs and HDIs.

The subject positionings are implicitly assumed – the reallocation is from ELUs to HDIs. The style is persuasive in nature and aims to convince the reader of the necessity of retaining ELUs as the only sensible thing to do. Persuasion in this context has nothing to do with showing any one person's beliefs to be inconsistent with certain absolute criteria of rationality, since that would reduce persuasion to a formal procedure operating within isolated and limited systems from which any uncertainty has been excluded. Persuasion, rather, takes the form of an attempt to make that person give up one set of beliefs in favour of another by offering a more or less thorough-going redescription of the world in which, on a pragmatic basis, a new set of beliefs is presented as more suitable, appropriate or probable (Rorty, 1989: 3–22, quoted in Torfing 1999: 67–68).

Existing lawful use

As noted in the previous chapter, the term existing lawful use 'came to mind' to one of the lawyers involved in the Water Law drafting team. Although the term was introduced in the Act, the policy process had the power to shape this idea: to strengthen it, weaken it or do away with it altogether. In the final version of the Water Allocation Reform strategy, Guideline 5 reads as follows: 'The water allocation process must be undertaken in a fair, reasonable and consistent manner and existing lawful uses must not be arbitrarily curtailed.'

Although texts are sites of struggle where the authors can no longer be identified, there are instances where the struggle is given character, and a particular controversy occurs between known actors. This was the case when the lead author of the NWRS, during an informal conversation at the national Water Summit on 5 May 2006, related how he and the DFID consultant had worked to convince the Director: Water Allocations of the need to stick to ELUs: 'With [regard to] existing lawful uses, [the DFID consultant] and I convinced him [the Director: Water Allocations] that he couldn't declare existing legal uses null and void. It would screw up the economy. Moreover, there was no provision in the Act for him to declare existing uses null and void.'

The Director: Water Allocation was also persuaded by the other two to tone down the 'take from the haves to give to the have-nots' policy, as they said that this would render the Department vulnerable to claims of arbitrariness. The use of rhetoric such as 'screwing up the economy' worked: their view prevailed and the notion of ELUs was retained in the WAR.

The early drafts of the position paper, though stating that the main purpose is redress, greatly emphasised the role of ELUs. In the draft dated January 2004, the section 'Principles to guide water allocation' contained no fewer than three principles dealing with ELUs, under the heading 'Promoting accountable and fair governance'. These principles were:

- Existing lawful uses will only be curtailed as a last resort and only after all other options to find water for the poor and BEE have been exhausted.
- Existing lawful uses of water will not be curtailed unless there are clear procedures and support programmes established to promote the productive use of water by emerging users.
- No existing lawful consumptive use of water will be completely curtailed.

In later versions, these three principles were condensed into one: 'It is critical to address equity needs, but attempts to deal with this must be balanced with the consideration that many existing lawful water users are making productive, efficient and beneficial use and are contributing to socio-economic stability and growth.'

The initial egalitarian emphasis – 'it is critical to address equity needs' – was qualified by the emphasis on a more utilitarian aspect in terms of water users 'contributing to economic stability and growth'. In the first draft of the paper, the idea of users having made 'productive and efficient use' was underscored by the following observation: 'Many existing users have made significant investments to make productive use of the water; irrigation is providing food security and contributing to the economy, while mining and industrial uses are providing employment opportunities.'

The emphasis on ELUs 'making investments', which draws on the Lockean notion of 'desert by labour' which, in rather oversimplified terms, holds that somebody who has worked for property should also obtain rights to that property, and downplays the historical acquisition of land. There was no hint of the dissent and debate that characterised the conception of the term ELUs. On the contrary, the policy discourse presented the retention of ELUs as the only 'natural' and 'pragmatic' thing to do. Dissent was reduced to consensus, and there was a reduction of possible options through discursive closure, that is, through marginalising other alternatives (Hajer 1995).

These passages serve to do three things: they construe particular subject 'positionings'; they privilege certain accounts over others by suppressing alternative readings about pollution and productivity (Freudenburg 2005); and they tie up the potential for reallocation with the productive capacity of emerging users, that is, they frame the problem of unequal water distribution mainly within the terms of making the most productive use of the resource in an equitable manner.

This section demonstrated how subject positionings were created in part through the coining of the term 'existing lawful use'. The following section deals with the category of HDIs, before discussing the notion of privileged accounts and the connection between productivity and equity.

Historically disadvantaged individuals

Social identity associated with the term ELUs was not only constructed through discourse. By implication, the identity of those who are not existing users was created through the highlighting of the negative relationship (a form of social antagonism): what one is (an existing lawful user), the other is not (Torfing 1999).

In the policy paper, HDIs[17] were construed as being more apt to create both economic disruption and environmental damage as they 'struggle to establish productive uses of the reallocated water' (this phrase was found in all versions of the paper, including the final draft). It is intriguing that the potential destructive capacity of 'emerging' users was commented upon, whilst neglecting to mention the problem of pollution, for example, from mines (see the following section). Existing users were portrayed here as 'beneficial', and the 'emerging' users – that is, HDIs – were portrayed as a potential threat to the environment. In what particular ways this 'environmental threat' might be manifested was not specified, and the reader is left to make his or her own conjectures. Consider the following excerpt from the first draft: 'If reallocations occur too quickly, the country will suffer economic and environmental damage as emerging users struggle to establish productive uses of the reallocated water.'

In particular, the second component of the argument, that 'environmental damage' will occur as 'emerging users struggle to establish productive use', implicitly draws on a larger meta-narrative of environmental myths (Adger, Benjaminsen et al. 2001; Forsyth 2003). In this case, the meta-narrative is the 'poverty–environment–degradation' hypothesis, a myth that has gained a strong hold on the imagination of both international development agencies and policy-makers (see, for example, Angelsen & Vainio 1998). There is an implicit assumption in the statement quoted from the first draft that reallocating to those who are not existing lawful users will result in environmental degradation, thus disregarding the potential environmental threat posed by the consumption levels of existing users (see, for example, Duraiappah 1998).

Most importantly, the narrative of ELUs firmly framed the problem of allocating water-use rights to HDIs on the basis of 'making efficient and productive use', which was emphasised through the phrase 'start people along the journey to becoming commercial and competitive users' that occurs in the final draft. Potential water users would be judged on their capacity to make productive use of water, which meant economically productive use. The narrative of 'existing lawful uses making productive and beneficial use of water' also posited putative causal links (Roe 1991) in that it predicted 'environmental degradation' and 'social instability' if water was taken away from ELUs. The narrative, or storyline, thus had a beginning, middle and end, and it foresaw given consequences of certain actions, providing a comprehensive interpretation of 'how things work' (Roe 1991; Fischer & Forester 1993).

Privileged accounts

The construing of existing users served to promote a discourse of existing uses – agriculture, mining and industry – as being singularly productive, thus making these sectors out to be purely beneficial in the sense that they provided employment and contributed to the economy. However, this story – what Freudenburg (2005: 91) would call a 'privileged account' – ignored the fact that both agriculture and mining were in decline and shedding jobs (Garduño & Hinsch 2005), and also ignored their environmental effects, especially those of mining, particularly through the phenomenon of Acid Mine Drainage (AMD). Freudenburg describes environmental damage as involving a 'double diversion', or two forms of privilege, where the first involves disproportionality (the privileged diversion of rights/resources) and the second revolves around the distraction of attention, largely through taken-for-granted 'privileged' accounts. Contrary to common assumption, much of the environmental damage is not economically necessary (Freudenburg 2005: 91), but results from privileged access to resources, such as through mining.

Mining has been a defining factor in South Africa's history and socio-economic fabric since gold was discovered in the Midrand in the 1880s (Ruiters 2002; Turton, Schultz et al. 2006), and was characterised by the exploitation of cheap migrant labour which laid the foundations for the wealth of the apartheid republic. Though BEE consortia have made some progress, Ruiters (2002) argues that the general composition of the mining sector has changed little since the days of apartheid, and that it is still characterised by stark inequalities.[18] Moreover, gold's fall from grace among many central banks has caused its economic position in the country to weaken (Ruiters 2002), and the industry has faced large retrenchments.

The extraction of gold and coal is extremely polluting and has a huge environmental impact, which has only recently come into the media limelight. According to Mariette Liefferink, the chief executive of the NGO Federation for a Sustainable Environment, 'In 120 years of gold mining, 120 gold mines have extracted 43 500 tons of gold and 73 000 tons of uranium, but they have left South Africa with a legacy of 410 km of tailings dams and 6 billion tons of iron pyrite' (Salgado 2009). This is particularly worrying since most mining activities in South Africa take place in the upper reaches of water catchments. The mining industry has been characterised by an enormous gap between rhetoric and actual mining practices (Ruiters 2002). This was evident in the Inkomati WMA, where farmers in the upper reaches of the Komati River worried about the potential spillage of toxic waste, such as sulphuric acid, from the Komati coal mine (farmer, Upper Komati, interview, 25 August 2006). There is little doubt that mining has contributed in part to creating the current water problems, for example, the water in the Olifants River Basin was considered to be of such poor quality that water had to be transferred from the Komati River Basin – some 130 m^3 each year – to provide Eskom's four coal-fired power plants with sufficient water of a good enough quality (Mpumalanga Department of Agriculture member of staff, interview, 23 June 2006). The detrimental effects of mining do not go away once the mine is closed down. On the contrary, the closure of mines represents a serious potential environmental hazard as mine water seeps into the ground and leads to underground aquifer contamination (Limpitlaw, Aken et al. 2005). AMD is a 'ticking bomb' according to some reports, which will affect the electricity and agricultural sectors particularly severely (Salgado 2009). 'AMD has been described as one of the most serious and potentially enduring environmental problems for the mining industry and, if left unchecked, could result in long-term water quality impacts that could well be the South African mining industry's most harmful legacy' (Prinsloo 2009).

Thus, the reform policy suppresses the notion that users may in fact be exploiting water in a polluting and wasteful manner (Hirsch 2005). Moreover, there is no mention of the fact that many catchments have been seriously modified by past productive activities, to which the following excerpt from Tlou et al. (2005: 2) attests:

> A number of water resource systems, the whole or parts of the resource, have been so seriously modified that the loss of natural habitat, biota and basic ecosystem functions is extensive (known as category E state). In this case, the resource base has been seriously decreased to the extent that rehabilitation to a functional ecosystem will be practically impossible.

Through framing the problem as dealing with naturalised scarcity, the point that existing users are part of the problem was neglected. Rather than explicitly acknowledging this, ELUs were presented as being unilaterally productive and beneficial. Though reference was made to the issues mentioned above in the NWRS and other documents, the WAR paper failed to mention it at all, thus creating a particular version of reality that did not include reference to the pollution risks associated with ELUs.

Efficiency and equity

Telling 'policy stories' and creating particular categories also creates the impression of particular causal relationships. Examining these relationships reveals something about the ways in which fundamental notions such as equity and efficiency are perceived. For example, in the January 2005 version, the following passage was found:

> it [is] recognised that the water allocation process, while providing water to the rural poor, should not fall into a 'sustainable poverty' trap,[19] which only provides water for small-scale livelihoods support, or for small-scale commercial gains. The water allocation process will therefore also seek to facilitate Black Economic Empowerment by promoting larger scale productive use by emerging users.

The use of 'trap' as a metaphor is a strong rhetorical device, rhetoric being, in the words of Watson (1995: 806) 'all about using language to persuade'. But not only that, it also shapes thoughts themselves. As McCloskey (1983 p. xvii, quoted in Watson 1995: 806) points out, 'figures of speech are not mere frills. They think for us.' The word 'trap' conjures up images of no escape. The juxtaposition with 'sustainable' is interesting in that it suggests that the 'trap' is linked with water, that is, that large amounts of water are needed. The term was later removed (after pressure from IWMI), but the emphasis on commercialisation was retained in the final passage (see below). In the same 2005 version, the following observation is made: 'It is not in the public interest to take water away from lawful, productive users before the emerging users have the capacity to use the water productively.'

In the March 2006 version, the passage as it reads above has been changed to finally read:

> In supporting the provision of water for uplifting the poor, the water allocation process must aim at providing water for subsistence purposes and sustaining basic livelihoods. Furthermore, it must also start people along the journey to becoming commercial and competitive users in their own right. The water allocation process must therefore support and facilitate Broad-based Black Economic Empowerment by encouraging and supporting larger-scale productive commercial uses of water wherever these opportunities exist.

Clearly, a concession has been made in terms of acknowledging water for subsistence needs as not necessarily representing a 'poverty trap'. However, this focus was qualified in the following sentence, which emphasised the idea of starting people 'along the journey to becoming commercial and competitive users in their own right'. There was no further mention in the document about subsistence, or small-scale, production.

What this did was to tie a tight knot between the constitutional claims of HDIs to greater access to water and their capacity for productive use. Greater equity could thus be achieved only through HDIs either gaining capacity to use water productively, or their getting access to benefits that accrue from water by being employed by someone who uses water productively. In essence, then, HDIs would get access to water on account of their productive capacity. However, this should not be applied to the detriment of ELUs, which were 'productive' and 'beneficial' by default. The subject positionings created in the policy narrative were thus linked to notions of efficiency and equity, where a more equitable distribution hinged on the capacity for efficiency, which was conceived as economic productivity.

Bourgeois theories of justice are of little use to the poor because they have scant opportunity to make use of litigation to seek atonement for injustices committed, argues Ruiters (2002). Although this is certainly often the case, teasing out the assumptions – or the 'prying apart' – of the policy is helpful in that it permits alternative frames to come into view, and it highlights how certain discourses and assumptions close down other avenues. For example, consider the following:

> equity in water use refers to equitable access to water, as well as actions to promote race and gender redress in water use. It is underpinned by the equitable distribution of the benefits of water use, and not the equal distribution of water.
> (Final version, November 2006).

Thus, the idea was not water resources *per se* that should be distributed equally, but rather the benefits accruing from the use of such resources. But how should such benefits be interpreted? One could claim with justification that all members of society in some way or other benefit from water use: through eating the food grown with irrigated water, by using the energy produced by power plants cooled by water, and so on. Therefore, what should such benefits be understood to comprise?

Basically, benefits in this context are understood to imply job creation. This is made clear in the concept of 'conditional licences' as described in the final position paper, that is, that licences will only be issued to those who can prove that they will provide employment to HDIs (DWAF Chief Co-ordinating Officer, interview, 20 July 2006). This created a dilemma that was not dealt with in any explicit manner, that is, how would considerations of balancing access to water itself and access to benefits accruing from water be achieved? In a scenario of compulsory licensing in a specific catchment, it might well be that not many applicants will come forward to apply for water. Should one then issue licences to companies that say they will provide jobs, even if the actual water resource is almost entirely in the hands of the former élite? Thus, could we envisage situations where a form of equity is achieved, where that equity is primarily measured in terms of 'benefits accruing from water use'?

The emphasis in the position paper on 'in the public interest' is key in this regard. The term is vague and difficult to interpret – the Water Law drafters were asked by Parliament to define 'public interest' but they were unable to arrive at a workable definition. 'Parliament asked us to define it, but we couldn't, so we left it out' (lawyer and member of the Water Law drafting team, interview, 10 August 2006). The idea of public interest, then, was very much open to contestations as the meaning of the term was defined and interpreted differently in different contexts by different people.

The style of the position paper suggests that the notion of 'public interest' was equated with that which promotes the greatest economic growth. However, it can be read as defending productive uses as these are in the public interest, that is, they contribute the most to the economy. Even where the paper does acknowledge that equitable distribution can lead to greater social stability (which is also in the public interest), it is not regarded as an end in itself, but rather as a means to promote further economic growth, which far more strongly invokes the principle of utilitarianism. For example, in the June 2004 version the following is found: 'this promotes social stability and hence investor confidence', and the January 2005 version contains the already cited phrase '...provides social stability, which in turn promotes economic growth'. Economic growth thus becomes the end rather than the means.[20]

Links to land: The absence of attention to acquisition

The NWA at no point actually mentions the word justice; instead, it talks about how to 'redress the results of past racial and gender discrimination'. A senior DWAF official noted that the previous Water Acts (the 1912 Irrigation Act and the 1956 Water Act) had not themselves been discriminatory in nature (interview, 5 May 2006). However, the riparian principle implied that land holdings determined water access, and the Acts dealing with land holdings, particularly the notorious 1913 Land Act, were certainly discriminatory. Thus, discussing water reform without making reference to what is going on in land reform would be fallacious.

A crucial issue in the land reform debates is that of compensation (Lahiff 2005; Ntsebeza 2007). Lahiff states that the payment of 'just and equitable compensation' is relatively clear from the Constitution, and also outlines an alternative or complementary approach: that of assessing land value according to productivity (production value). According to Ntsebeza (2007), Chaskalson (1993, cited in Ntsebeza 2007), optimistically interpreted the property clause to imply that expropriation did not necessarily have to take into account the market value of the property to be expropriated. However, the Gildenhuys formula, conceived by Judge Gildenhuys in the early phases of the land reform process that got underway after the transition, provided a guideline to determine the compensation to be paid in terms of 'the market value minus the current value of past subsidies'. Although the inclusion of subsidies reflects the extent of State support lavished on white farmers during the apartheid era, Ntsebeza notes that it is

> intriguing that the history of how colonialists acquired land in the first place is not receiving prominence in the discussion of compensation. In so far as reference is made to history, the suggestion is that this refers to the history of land acquisition by the affected landowner. Yet, there is the history of colonial conquest and dispossession that lies at the heart of the land question in South Africa.
> (Ntsebeza 2007: 124)

Ntsebeza's point is that the land reform tends to pay comparatively little attention to the nature of the resource capture itself, and thus fails to address the question of justice of acquisition and how property came to be possessed in the first place. Ntsebeza also notes the absence of any discussion on how the 'naked exploitation' of African labour contributed to the success of white commercial farmers (see Ntsebeza 2007: Chapter 3). This echoes Wolpe's observation that while the two sectors (commercial and subsistence agriculture) were presented by past governments as reflecting 'modernity' and 'tradition', respectively, the economic function of the black 'reserves' was to reproduce, and subsidise the cost of, labour (1972, cited in Hall 2004: 213).

The land reform of South Africa has undergone a marked shift in its nature, from an initial emphasis on rural livelihoods to focusing on commercial farmers, and the restriction of ambitions to a 'limited deracialisation of the commercial farming areas rather than a process of agrarian restructuring' argues Hall (2004: 213). In a political economy context, this shift is 'consistent with changes in macro-economic policy and reflects shifting class alliances' (Hall 2004: 213).

Narrowing down the 'room for manoeuvre'

What has happened within the drafting of water legislation is that the room for manoeuvre provided for by the 'safeguard' clause has been narrowed down by the continued focus on the need to protect existing property rights and discursively construing users as 'productive and beneficial'. Not only is the principle of protection of private property invoked, but it is also framed as being in the public interest that existing users maintain rights, in order not to 'destabilise the economy' or trigger 'environmental degradation'.

The idea that individuals who were using water under the auspices of the old 1956 Act have been declared existing lawful users resulted in a profusion of legal suits against the government, with reference to section 25 of the Constitution, known as the 'property clause'. According to the DFID consultant,

> What the Act has done is to create a very technically complex material, which has created opportunities for the right to appeal by taking the hard line in terms of saying 'we are going to take away your water, but we are not going to pay you any compensation'. DWAF is now losing most of the court cases, it's costing a lot of money, and the Department is losing face over this. It would be easier and probably cheaper to say that 'we are going to pay compensation where there is severe economic prejudice' and just leave it at that.
> (Interview, 12 May 2006).

There is a complete absence of discussion around colonial appropriation of water, or the 'justice of acquisition', as Ntsebeza (2007) notes, for land.

However, distribution of resources according to certain criteria, or what Nozick (1974) calls 'patterned distribution', calls for continuous intervention in people's lives and violates their fundamental rights. A distribution is just if acquisitions and transfer of holdings are just, argues Nozick, thus dismissing ideas of redistribution *except* when the aim is the rectification of past injustices in the acquisition of holdings. In that case, 'if past injustice has shaped present holdings in various ways, some identifiable and some not, what, if anything, ought to be done to rectify this injustice?' (Nozick 1974: 152). What obligations do the agents of injustice have towards those whose position has been worsened by the injustice? In addition, how, if at all, do things change if the beneficiaries and those made worse off are not the direct parties in the act of injustice, but their descendants? There is no thorough or theoretically sophisticated treatment of such issues, argues Nozick (1974).

However, conceiving of justice as a pluralistic concept (Munzer 1990; Sen 2009) allows for the teasing out of underlying assumptions that are at play in justifying certain patterns of property rather than others to manifest themselves. With respect to water rights, the specific characteristics of water as constituting a complex ecological system pose even greater challenges, not least in the assumptions inherent in the belief that rights can be precisely quantified in a meaningful way. I thus argue that the construction of water rights draws on ideas about putative merit (in using a scarce resource in a productive manner) that subsumes the larger questions about distributive justice.

This chapter has demonstrated how the national-level policy narrative centred on scarcity, and how it created certain subject positions and vested important concepts with particular meanings. It highlighted how the emphasis on retaining ELUs narrowed down the room for manoeuvre, whilst the notion of emerging users becoming commercial and competitive water users in their own right was emphasised. How do these ideas and positions mesh with local-level discourses and understandings? The following chapter focuses on discourses and practices in the Inkomati WMA.

Notes

1. The full title of the book was *An essay on the principle of population as it affects the future improvement of society, with remarks on the speculations of Mr. Godwin, M. Condorcet, and other writers*. Malthus's argument was rather crude, and failed to acknowledge Condorcet's more subtle line of reasoning. See Sen (2009) for a discussion of Condorcet's views.
2. As an example of the importance of adaptive capacity, consider the cases of a water-abundant country such as Canada, which is prone to frequent water shortages in terms of errant service supplies, and Barbados, with comparatively limited amounts of water, which is coping rather well (Wolfe & Brooks 2003).
3. Another estimate of water per capita puts it even lower: 'With an annual mean run-off of 49 228 million m^3 and a total population of 44.8 million, the per capita water availability works out to be 1 099 m^3/year. This is only marginally above the level of 1 000 m^3/year considered to indicate a state of water stress.'
4. Government Notice 1352 of 12 November 1999.
5. Nationwide, the total estimated volume in the NWRS was 12.9 billion m^3/year, whereas the registered volume was 16.9 billion m^3/year.

6   The Water Research Commission (WRC) had commissioned research on the potential of developing more dynamic approaches to allocation than those currently allowed for through the old yield models used by DWAF. These yield models harked back to the original European-imported Water Resources Yield Model (WRYM) that was used in the 1950s and 1960s. They were not really suitable for South African conditions, because of the complexity of water resource networks in the country and the extreme climatic conditions. However, the imported model was adopted by DWAF, as it had a very flexible structure that could be upgraded and modified. The basic principle remained: estimating a system's yield. In April 2006, a workshop was convened by the WRC in Richards Bay to discuss the results of the commissioned research. A more dynamic system emerged: Fractional Water Allocation and Reservoir Capacity Sharing (FWARCS). It was reasoned to be considerably more flexible than the yield modelling approach, as it allowed for users to 'bank' the water they did not use, and thus did away with the 'use-it-or-lose-it' approach that prevailed (Pott A, Hallowes J et al. (2005). A computerised system was developed for auditing real time or historical water use from large reservoirs in order to promote the efficiency of water use (*WRC Report* No. 1300/1/05. Pretoria: Water Research Commission). However, resistance at the time was fierce, with sceptics observing that the model was 'far too sophisticated for South African conditions' (workshop participant, interview, 25 April 2006), and dismissing the 'sales talk' that the fractional modellers engaged in, saying that there was a kind of turf war going on within DWAF with respect to these approaches. The FWARCS model is, however, used to determine international water sharing arrangements between South Africa, Swaziland and Mozambique (see Dhlamini et al. 2007).
7   The research themes that underpin the CSIR's core research activities in this domain include coupled land, water and marine ecosystems, energy futures, environmental assessment and management, forestry resources, mineral resources, pollution and waste, sustainability science and water futures (see www.csir.co.za).
8   See also www.csir.co.za.
9   The Water Research Act (No. 34 of 1971) established the WRC in order to promote research in connection with water resources.
10 See www.iwmi.cgiar.org/africa/south.
11 An indaba, a Zulu word meaning 'business' or 'gathering', is an important conference held by the *izinDuna* (principal men) of the Zulu and Xhosa peoples of South Africa. Such indabas may include only the *izinDuna* of a particular community or may be held with representatives of other communities. The term is widely used throughout Southern Africa, often simply meaning gathering or meeting.
12 This could be changing. At least there are now efforts to link the processes better, such as the national Water and Agrarian Reform workshop, held in Pretoria in March 2009, which aimed to couple the land and water reform processes more tightly.
13 Although there is available documentation on customary laws elsewhere in Africa (see Chikhozho & Latham 2005).
14 Of this percentage, DWAF estimates that irrigated agriculture makes up a mere 25–30 per cent. However, these estimates do not take account of the forward and backward linkages of agriculture with other economic sectors.
15 The full document can be accessed at www.dwa.gov.za.

16 Although the term is an attempt to avoid the old regimes' crude classification according to colour, it is still problematic in that it designates a group in line with a vague sense of historical disadvantage, and thus includes anyone who could have been said to suffer under the previous regime. The problem with such a classification is only too clear in the increasing allegations of 'capitalist cronyism' levelled at those who have benefited from affirmative action programmes such as the BEE strategy. It fails to adequately distinguish those in real need of affirmation and empowerment, and voices have been raised to suggest that perhaps it would be better to devise some sort of criteria according to income levels, to avoid the sort of class gap that is now becoming more and more evident in the modern South Africa (see Alexander 2006).
17 Historically disadvantaged individuals are defined in Preferential Procurement Policy Framework Act (No. 5 of 2000) as a South African citizen
   i) who, due to the apartheid policy that had been in place, had no franchise in national elections, prior to the introduction of the Constitution of the Republic of South Africa, 1983 (Act No. 110 of 1983) or the Constitution of the Republic of South Africa, 1993 (Act No. 200 of 1993) (the 'Interim Constitution'); and/or
   ii) who is a female; and/or
   iii) who has a disability.
18 See also http://www.queensu.ca/samp/migdocs/Documents/2000/NUM.htm
19 The 'poverty trap' thesis specifies a circular or spiral relationship between poverty and environmental degradation, that is, environmental degradation and poverty reinforce each other: the poor are both agents and victims of environmental destruction (see Angelsen & Vainio 1998).
20 In the final version this is toned down somewhat to read 'Job creation and economic growth are among South Africa's most important priorities.'

4

Water allocation in the Inkomati

THIS CHAPTER EXAMINES WATER allocation issues in the Inkomati Water Management Area, one of the pilot sites where water allocation reform was to be implemented. The different perceptions and ongoing struggles over authority are teased out, based on interviews with individuals and groups of existing and potential water users and officials within relevant departments. The local understandings and discourses on water use and users, allocation and access are contrasted with the policy-level views and discourse. The chapter ends by giving an account of the preparations for compulsory licensing, and the resulting outcomes.

Context: Into the Inkomati

The Inkomati is one of two Water Management Areas[1] (WMAs) to be found in Mpumalanga Province. Mpumalanga, which means 'where the sun rises' in Zulu, was known as the Eastern Transvaal under the apartheid regime. The Inkomati, which includes the southern parts of the Kruger National Park, shares its south-eastern border with Swaziland, with Mozambique lying to the east (see Figure 4.2). One needs to understand the socio-historical context in order to grasp the constellation of interests and discourses in the area. The following sections provide an overview of the region, its history, its characteristics, the water users, infrastructure and institutions present, and highlight how the historical legacies of colonialism and apartheid have had lasting effects on the nature of present water use patterns, and how these have shaped the current discourses and practices with respect to water and water allocation reform.

Historical legacies shaping patterns of water use

Settling the Transvaal

British domination in the Cape Colony prompted many of the Boers to move inland in the 1830s, a migration that became known as the Great Trek. The Boers, or Afrikaners as they also called themselves, strongly disliked being under British rule. They were strong spirited, protective of their individual freedom and distrusted collective institutions (Sparks 2003b). The coming agonies of the Anglo-Boer War and the Great Depression were to give rise to a strong sense of Afrikaner nationalism, which 'shaped and drove South Africa as a malign force for nearly the whole of the twentieth century' (Sparks 2003b: 5). The history and self-image of the Trekboers were shaped by their desire for self-sufficiency, their strong-headedness, and being deeply rooted to the land. Courting images of fertility and 'making the desert bloom', the Boers came to view themselves as the 'chosen people', and the land they occupied as God-given, drawing upon Jewish images of Zionism (Sparks 2003b, also interview with CSIR representative, 2 May 2006).

The *Difaqane*, having wrought havoc in the area (see Chapter 2: 12–13), enabled the Boers to settle large swathes of land without too much trouble, and the subsequent insidious legalisation of this massive land grab through a string of discriminatory Acts cemented their power base in the Transvaal. The Glen Grey Act of 1894 (Terreblanche 2002) set a limit to land ownership in black reserves, and this gradual dispossession of blacks culminated in the passing of the notorious Natives Land Act (No. 27 of 1913) and the Bantu Trust and Land Act (No. 18 of 1936). These pieces of legislation combined to essentially reserve 13 per cent of the land for people of African descent, who at that time made up 70 per cent of the population (Bate & Tren 2002: 60). Blacks were banned from using tractors in 1936 and, in 1952, irrigation was heavily restricted. The effect was to deny self-sufficiency to blacks, and the self-determination promised by the apartheid advocates never fully materialised (Schreuder 1994, in Bate & Tren 2002). Building on the indirect rule of the British, the Nationalists supported the chief and tribal system as a method of control (Mamdani 1996).

The discovery of gold was what initially drew the English and Afrikaners, as well as the prospect of trading with the coastal areas. Later, however, the increasing awareness of the agricultural potential of the region, particularly the south-eastern part of the Transvaal – the low-lying, fertile Lowveld – became the primary reason for settling there. But the only way that farming could develop, according to a 1952 report by the Irrigation Finance Commission, was if the utilisation of water supplies was controlled entirely by the State, since 'private enterprise cannot provide the money necessary to finance them' (Commission 1952: 6, quoted in Bate & Tren 2002: 140). This marked a high point in the era of supply-side management as the State embarked on its 'hydraulic mission' of building dams and infrastructure to support the white farmers across the country (Turton & Meissner 2002).

The 1956 Water Act facilitated the proclamation of a Government Water Control Area (GWCA) in order to improve the distribution and utilisation of water resources. In 1971, a GWCA was proclaimed[2] on the Komati River and its tributaries, a key water source in the Lowveld. Ten years later, the government decided to withdraw its control of the tributaries, and merely concentrate on the main river (Waalewijn 2002). The GWCA was established primarily as a result of the political and economic importance of the region. According to Bate and Tren (2002) it was politically important as the farmers of the Eastern Transvaal represented a significant constituency of votes, and economically important because of the revenues brought into State coffers by the citrus crop exports. Hence, the farmers were provided with ready access to support and subsidies (Deacon 1997, cited in Bate & Tren 2002: 140) and had access to cheap African labour (Marais 2001; Terreblanche 2002).

Sweet dreams: The growth of sugar as a political force

Sugar had been produced primarily by English-speaking farmers in the Natal area, but once the first sugar mill was built at Malelane in the lower Komati catchment in 1965, sugar gained importance as one of the major crops in this area too. Sugar growing was significantly supported by legislation, such as the Sugar Act (No. 28 of 1936), which was promulgated to provide protection for farmers during the Depression when the price of sugar slumped world-wide. Although amended several times, the Act remains in force with its substantive contents intact. The legislation centralises control of the sugar industry by specifying national and export quotas to be produced. Fixed prices and a ready market renders sugar farming a far more reliable and relatively profitable activity in contrast to more risky ventures, such as vegetables and citrus, which are subject to price oscillations and slumps in demand (Martin Slabbert of TSB Sugar, interview 8 June 2006).

With the National Party's accession to power in 1948, the concomitant emergence of the Broederbond (see Chapter 2: 16) quickly became a social force to be reckoned with. Braam Raubenheimer, who served as Minister of Water Affairs in the 1970s, and who was also a Broederbond member, lobbied intensely to secure positions for Afrikaners only at all levels of social and administrative life, in order to 'neutralise the oppression from the English-speaking' (Bate & Tren 2002: 144). State support for farmers was an important component of this policy. The company Transvaal Suiker Beperk (TSB), later re-branded TSB Sugar, was created in 1965, following the establishment of the sugar mill at Malelane on the Lomati River, a tributary of the Komati. The Broederbond was a major shareholder in TSB, and was therefore in an 'excellent position to expand production, lobby for increased water allocation and represent the interests of the sugarcane farmers upon whom they relied' (Bate & Tren 2002: 145). It was argued that the Broederbond used the economic importance of sugarcane as a political weapon, and the area was 'frequently favoured in the granting of increased abstraction rights in addition to the normal flow permits and was particularly favoured in times of drought' (Bate & Tren 2002: 145). Complementing the liberal granting of water abstraction rights, the farmers received generous support from the government, both in terms of infrastructure development and management (dams, canals and weirs) and direct subsidies, which afforded them enormous benefits. No wonder the area was a National Party stronghold until 1994.

Creating the KaNgwane homeland

Three years after the National Party came to power, the Bantu Authorities Act was passed, emphasising the apartheid strategy of physical separation, with the ultimate aim of all homelands becoming independent (see Chapter 2: 16). Thus started twenty years of forced removals, which uprooted more than 3.5 million people, in a massive social engineering effort to realise the Boers' ideas of 'separate development' (Marais 2001; Terreblanche 2002; Sparks 2003b).

Intrinsic to the idea of apartheid was the notion that all 'natives' should have their own homelands, or 'self-governing' territories. Whilst South Africa's 10 homelands were considered to be separate political spheres that were progressively working towards independence, it was the Department of Native Affairs (DNA) that was responsible for their running (Ashfort 1997 & Evans 1997, cited in Oomen 2005: 18–19). The proliferation of regulations and bye-laws produced by the DNA was based loosely on ethnographic data 'derived from encounters between native administrators-cum-state anthropologists and traditional leaders and village elders' (Hammond-Tooke 1997, cited in Oomen 2005: 19). As with indirect rule (see Chapter 2: 16) traditional leaders came to be taken as the representatives of their communities, and those not amenable to their new role as a bureaucratised institution were dispatched to re-education camps or simply removed (Oomen 2005).

One such homeland, KaNgwane, was demarcated in the Lowveld area by the border of the Komati River, and parts of it today make up the Nkomazi area. In the nineteenth century, the bakaNgwane people occupied what was to become the present-day Swaziland. When Swaziland was declared a British High Commission Territory in 1907 and the land partitioned, many of its people were left outside and thus targeted for the creation of a separate homeland.

FIGURE 4.1 *Map of Nkomazi area*

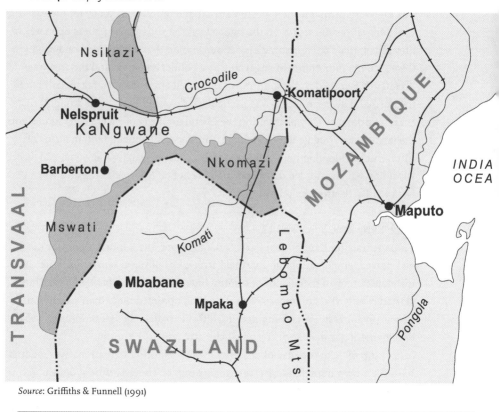

Source: Griffiths & Funnell (1991)

The bakaNgwane did not live in a single area but were scattered in several places as more or less distinct communities. Thus, the 1975 homeland consolidation proposal resulted in the homeland comprising two tracts of land contiguous to Swaziland – Nkomazi and the rather barren Highveld area of Mswati – as well as the detached Nsikazi. Presumably, the idea was to remove the people of Nsikazi to the Nkomazi/Mswati area later. KaNgwane was formally proclaimed a 'self-governing territory' in 1977, with a total area of 385 000 ha on which lived roughly 1 million people. KaNgwane thus became the formal apartheid home of the 'leftover' SiSwati-speaking bakAngwane people until it was formally dissolved in 1994 with the transition to non-racial democracy.

The areas remained segregated, with the former homeland subject to the 'overcrowding, corruption, migrant labour and systematic underdevelopment' that were common conditions in many of these homeland areas (Nxumalo 1997: 4) while the large, white-owned farms exploited the more fertile land to the north. This resulted in widespread exploitation, depressed wages, and 'shocking health conditions in a group that tries to remain invisible' (Nxumalo 1997: 3). Most people in the area are Swazis, who subscribe to both traditional and Western cultural and religious practices – Christianity thrives amidst strong support for traditional healers. Far from being a cohesive community, however, the area is characterised by social tensions, particularly in terms of immigrants flowing into the area from across the rather porous border, who were seeking refuge from political instability, and were attracted by the income opportunities on the farms (Polzer 2007).[3] During an interview on 25 May 2006, a local, white Irrigation Board chairperson described the tensions between the migrant Mozambicans and the local Swazis as follows:

> You won't hear Shangaan spoken; it is mostly Swazi that is spoken in this area, although the Shangaans are in majority. This is because they don't want to be seen as outsiders or intruders. There is a lot of slandering among groups and outsiders are perceived as enemies.

There were clear signs of hostility towards Mozambicans as a group. One dryland farmer, present at a group meeting, complained that they had had to move their cattle further away from their homes, and to send herders with them, because 'the Mozambicans are stealing our cattle' (interview, 20 June 2006). The history of forced removals had also contributed to tensions in the area, as the resettlement of people of the Hoyi, Siboshwa and Lugedlane tribes from the Tenbosch area (south-east of the Kruger Park) to the KaNgwane had created 'much local tension as they were settled on top of the land rights of the existing community in the area. This has created a situation of conflicting and overlapping rights on the same area of land and has created volatile living conditions for the population' (PLAAS 2004: 1).

In terms of current population figures, the sources tend to be fraught with large degrees of uncertainty, not least due to the shifting of boundaries between the 1991 and 2001 censuses. The 2001 census for Nkomazi puts the total population in the area at 334 416. Fifty-three per cent of the resident population is female, and close to half – 41 per cent – is 14 years old or younger. Only 4 per cent of the population is 65 years or older. The area has extremely poor standards of educational achievement and poor facilities, typical of a homeland area. Of those over 20 years of age, 29 per cent do not have any schooling at all, 23 per cent have completed secondary school, and only 5 per cent have any education higher than grade 12 (Stats SA, Census 2011).

Sharing waters: The 1992 Agreements and the forging of new boundaries

Irrigation development in the KaNgwane – later the Nkomazi – started in 1985, with crops such as cotton, maize, vegetables, leather fern and sisal and some sugarcane. Before this, the area was mainly used for cattle farming and dryland agriculture. There were also vegetable gardens for community use in order to foster food self-sufficiency (Head of Technical Support Services at DALA, interview, 23 June 2006).

It was decided to build two new dams in 1992, the Maguga and the Driekoppies, on the Komati–Lomati system; 'something we had been fighting for all along', according to a former Irrigation Board chairperson (interview, 2 August 2006). Swaziland and South Africa negotiated an agreement on water sharing, which gave details of how water from the Komati River Basin was to be allocated. During this process, the Joint Water Commission (JWC), a controlling body for all shared watercourses between South Africa and Swaziland, was established. In addition, the Komati Basin Water Authority (KOBWA) was created as an independent body with the purpose of overseeing the operation and management of the dams. However, KOBWA took on a much broader role than mere oversight. Among other things, it was devising management plans for the Komati and Lomati rivers, as well as determining in-flow stream requirements (Enoch Dhlamini, interview, 3 August 2006). At the same time as the Treaty was negotiated with Swaziland, South Africa forged a sharing arrangement with the KaNgwane homeland. According to the treaty, the KaNgwane homeland was allotted 120 million m³ of low assurance[4] water for irrigation purposes. (see Table 4.1)

South Africa had almost 140 million m³ of high assurance water, which it used for primary and 'strategic' purposes, including meeting the water needs of the electricity-generating parastatal Eskom. Eskom was defined as a 'strategic user' and therefore needed water supplies with a high level of assurance. KaNgwane was given 6.6 million m³ of high assurance water for primary purposes, which was over and above the 120 million it got for irrigation – although only 95 out of the 120 million m³ were actually allocated to irrigation, the rest being retained for

TABLE 4.1 *1992 water sharing treaty: Allocation of water between the KaNgwane and the RSA*

Water consumptions (million m³)	High assurance	Low assurance	Totals
KaNgwane	6.6	120.4	127
South Africa	139.8	252.2	392

future development. The KaNgwane homeland administration distributed the 95 million m³ amongst the seven tribal authorities that made up KaNgwane, 'taking into account population and geography...[and trying] to be fair to everyone' (Head of Technical Support Services at DALA, interview, 23 June 2006). The remaining unallocated portion, however, was to become a bone of contention after the demise of the homeland administration.

Boundaries were redrawn shortly after the transition in 1994 to correspond with the vision of a new South Africa and a radical new Constitution. This represented a major upheaval of the existing lines of division, with the former regions being morphed into nine new provinces. Bringing the former homelands into the fold of the new Rainbow Nation was particularly challenging – the former homeland administrations had to be absorbed into the new systems of governance and new alliances forged between tribal authorities and structures of local government, with the latter being instituted 'wall-to-wall' across the country and provided with extended powers.

Traditional leaders managed to assert considerable authority for themselves in the new Constitution. The ANC was keen to ensure that they were included in the fold after the end of apartheid, as it was anxious to avoid a split between its urban power base and some of the more tradition-oriented parts of the constituency in the countryside, which remained firmly embedded in systems of chieftaincy. This anxiety on the part of the ANC resulted in the creation of the Traditional Leadership and Governance Framework Act (No. 41 of 2003) (TLGFA), which created space for the traditional authorities to exercise power through, for instance, providing for the establishment and recognition of traditional leaders and councils (see Chapter 2: 25).

There was an uneasy co-habitation of the liberal democratic values that the ANC preached and the more authoritarian and patriarchal conventions that are imbued in the system of chieftaincy. The practice of indirect rule brought in chiefs who were amenable to the system and who were more interested in securing their own powers than in furthering the interests of their communities. More recently, the process of trying to enact the Communal Land Rights Act (CLaRA) has brought to the fore the schism between 'traditional' and liberal values. The Act vested much authority in chiefs which, argue Claassens and Cousins, among others, was to the detriment of women (Claassens & Cousins 2008: see also Chapter 2: 25).

In Mpumalanga, the traditional systems were still very much alive in terms of the power that chiefs exercised over land allocations, as documented by, for example, Levin and Mkhabela (1997). The ANC's vacillation (Lahiff 2003; see also Oomen 2005) on the question of traditional authority contrasted sharply with the opinion of civic representatives in the area, who maintained that the chieftaincy had become increasingly irrelevant and would have withered away had it not been for the ANC's continued indeterminism on the issue.

Characteristics of the Inkomati Water Management Area

When the Water Act was passed in 1998, the country was demarcated in yet another way – along the hydrological boundaries of catchments. This demarcation exercise gave rise to 19 Water Management Areas, of which the Inkomati was one.

Geography and water availability

The Maputo-bound N4 highway from Johannesburg first traverses miles of flat Highveld plains at about 2 000 m above sea level, before descending from the Great Escarpment of the Transvaal and the Drakensberg into the Lowveld region (about 140 to 800 m above sea level). The shift is gradual, from the golden flat horizons and crisp air of the plateau, with its crops of wheat and pasture, through the rather steeply-sloping gradient dominated by miles and miles of pine and eucalyptus plantations, and then down into the sunny and warm Lowveld with its subtropical crops of sugar, citrus, bananas, nuts, and vegetables. Rainfall is strongly seasonal, and the entire basin falls within the summer rainfall region, concentrated in the summer months of December to March. Mean annual rainfall varies from more than 1 200 mm per annum in the mountains and on the plateau in the west, to as low as 400 mm in the eastern Lowveld area (DWAF 2004a).

Three independent catchments make up the WMA: the Crocodile, the Sabie-Sand, and the Komati. The Komati and Crocodile rivers both originate on the plateau and then flow downwards to feed the Lowveld, which has 'some of the most productive and valuable agricultural land in South Africa' (Bate & Tren 2002: 135). The Sabie-Sand River system extends into the Kruger Park and remains the most pristine catchment. All three catchments drain eastwards and ultimately flow together, crossing into Mozambique to become the Incomati River, and eventually empty into the Indian Ocean. Because the Komati River originates in South Africa, then crosses into Swaziland and re-emerges in South Africa again before the confluence with the Crocodile River on the Mozambican border, South Africa becomes both an upstream (west of Swaziland, west of Mozambique) and downstream (north of Swaziland) country in this region.

Grains and vegetables are grown on the high plateau, whereas the Lowveld is more amenable to intensive commercial agriculture, including crops such as sugar, citrus, subtropical fruit, and horticulture (Bate & Tren 2002). Commercial agriculture has been accompanied by the intensive development of surface storage for runoff, much of which was privately owned under the previous legislation. The bulk of agricultural development occurred in South Africa. Development in the homelands of KaNgwane, Lebowa and Gazankulu were reportedly constrained by traditional land tenure arrangements, as well as a lack of access to capital and extension services, but also due to marginal and low-potential soils (Woodhouse 1997; Brown & Woodhouse 2004). Apart from the irrigation schemes initiated by the government, as described above, most farming was with dryland crops, typically cotton and maize.

The Crocodile River has only one major dam, the Kwena, located in its upper reaches (although it does have a number of weirs and diversions). The Komati River has been much more intensely developed, with two dams, the Vygeboom and the Nooitgedacht, being built in the 1970s. These dams were built to ensure a stable flow of water into the neighbouring Olifants River catchment for the purpose of power generation. In addition, the Komati River has a number of weirs, originally built by the settlers, which have a total estimated capacity of 19 million m³ (MBB 2000, quoted in Waalewijn 2002). The Lomati River has only one weir. The weirs act to stabilise the river flow, and are owned and operated by the Irrigation Boards in the Lower Komati, but many of these were destroyed in the floods of 2000. There is also the Masibekela Dam, an off-channel storage dam with a capacity of 9.1 million m³ for storing floodwater from the Komati River. Some years ago, work was completed on two further large dams. The Driekoppies (or Matsamo) on the Lomati River on the border with Swaziland, was finalised in 1997 to the tune of R600 million, and has a capacity of 251 million m³. The 332 million m³ Maguga dam on the Komati River, within the borders of Swaziland, was completed in 2003 at a cost of R800 million.

Irrigated agriculture is by far the major water user in the Inkomati, accounting for an estimated 565 million m³ of water use in total per annum (DWAF 2004a). Other users include forestry, mining, recreation, and ecotourism – the Kruger National Park is a major attraction. Forestry, mainly found in the escarpment areas where the soil is too thin to till, is also a big water user. Intensive forestry has a considerable impact on stream runoff, which is why in the National Water Act, stream flow reduction activity is portrayed as a new category of water use (see Chapter 2). The foresters themselves complain about this, calling the water use charges levied on them a 'rainfall tax'. Vast stretches of monotonous forestry plantations, consisting mostly of the fast-growing eucalyptus and pine species, dominate the sloping hills that mark the transition from the Highveld to the Lowveld. The trees are used mostly for pit props and for pulp production. It is essential that the trees are as straight and fast-growing as possible for pit props, which means that they are very 'thirsty' trees. In the Inkomati, the industry is dominated by two major players, Sappi and Mondi, and the total afforested area in the Inkomati WMA has been estimated to be approximately 3 357 km² (Brown & Woodhouse 2004). Mining is not extensive in the area, but represents an increasing environmental risk, since the mines are located in the upper reaches of the catchment (Chairperson, Upper Komati Water User Association, interview, 25 August; also DWAF 2004a: iii). More recently, concern is growing that the problem of Acid Mine Drainage (AMD) will become a problem in this area as well, as it has in the Vaal catchment (Salgado 2009).

FIGURE 4.2 *Inkomati Water Management Area (WMA)*

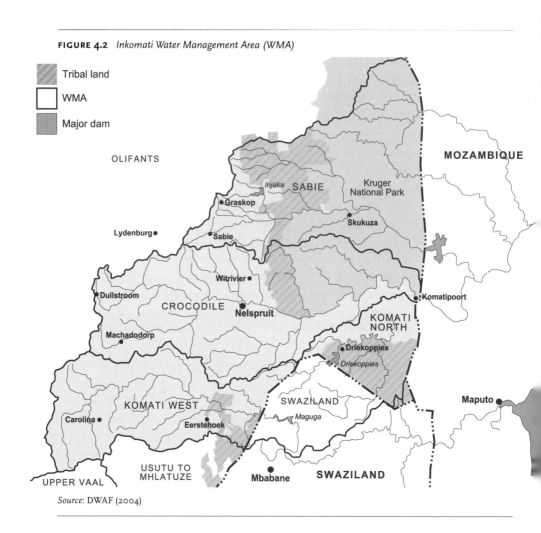

Source: DWAF (2004)

TABLE 4.2 *Water yield and demand, NWRS 2004*

Sub-area	Estimated local yield	Local demand	Balance
Lower Komati	252	292	−40
Upper Komati	118	162	−44
Crocodile	258	413	−155
Sabie Sand	95	117	−22

Source: 2000 situation, adapted from NWRS 2004 (DWAF 2004b).

Notes: All figures are in million cubic metres per year.
Swaziland, and details of transfers in/and out of basin, are not included here.

TABLE 4.3 *Water yield and demand, ISP 2004*

Sub-area	Estimated local yield	Local demand		Balance
Lower Komati	298	304		−6
		Irrigation	222	
		Int'l requirements	60	
		Forestry	12	
		Rural	6	
		Urban	3	
		Industrial/mining	1	
Upper Komati	118	159		−41
		Eskom	109*	
		Forestry	23	
		Irrigation	21	
		Rural	4	
		Urban	2	
Crocodile	264	413		−149
		Irrigation	257	
		Int'l requirements	49	
		Forestry	42	
		Rural	7	
		Urban	35	
		Industrial/mining	23	
Sabie Sand	149	111		+38
		Irrigation	54	
		Forestry	34	
		Rural	2	
		Urban	13	
		Transfer out	8	

Source: Adapted from ISP 2004 (DWAF 2004a).

Notes: 2004 situation, with figures broken down into sub-sectors.

All figures are in million cubic metres per year.

* The NWRS reserves this transfer (i.e. requires national authorisation) up to 132 million m³ per annum, even though the current capacity of the system is only about 97 million m³ per annum. The implication of this is that transfers out of this sub-area could increase in future, although new infrastructure would have to be constructed in order to achieve this. The yield of the Vygeboom and Nooitgedacht dams is insufficient to transfer as much as 132 million m³ per annum, especially after making allowance for the ecological Reserve. Department of Water Affairs and Forestry (2004a) *WMA05 Inkomati: Internal Strategic Perspective*, Pretoria: 32.

Tables 4.2 and 4.3 show the estimated yield and demand in the area, categorised according to sub-catchments, with the Komati catchment divided into the Upper (west of Swaziland) and Lower (north of Swaziland) catchments, respectively. The first table draws on the National Water Resources Strategy (NWRS) (Department of Water Affairs and Forestry 2004), while the second table is taken from the Internal Strategic Perspective (ISP).

There was a considerable degree of discrepancy between the sources on water availability. The NWRS estimate of water availability in the Inkomati region dated from 2000, when the Maguga Dam had yet to be filled. Added to these data were high-demand and low-demand scenarios, where demands were extrapolated until 2025.

The ISP, however, depended on data gathered in 2003 and factored in international requirements.[5] It also took account of the extra yield made available through the filling of the Maguga Dam, but noted that the building of the dam had increased the impact that forestry would have on the system. Unlike the NWRS, the ISP also treated the Sand River catchment as a separate sub-area on account of it being 'different to [sic] the Sabie catchment from a water resources perspective' (Department of Water Affairs and Forestry 2004a: 54). The ISP also argued that the groundwater estimates provided in the NWRS 'seemed very low'. For example, it quoted groundwater use as 2 million m³ in the Lower Komati catchment, whereas the registrations in the WARMS listed groundwater use as 9 million m³, mainly for irrigation. Garduño and Hinsch (2005: 52) suggest that countrywide there is roughly about 50 per cent more groundwater available than was reflected in the NWRS. The NWRS was also inconsistent with regard to the reduction in yield due to forestry in the Upper Komati area. Added to these observations was the fact that the Ecological Reserve had yet to be defined, much less implemented, in the region.

These discrepancies in water yield estimates highlight the uncertainties inherent in modelling approaches. The complexities associated with the interaction between stream flow reduction activities and storage are merely hinted at. This underscores how perceptions of scarcity depend on the available modelling approaches and technologies, and how the choice of approach contributes to construing a particular perception of a given situation. It highlights the futility of the search for 'correct' answers, as what is perceived to be 'correct' depends on the assumptions and criteria associated with a particular model and the people operating it. Acknowledging uncertainty and operating with ranges rather than discrete numbers would probably be more prudent.

Water management structures

A new three-tier structure of governance took shape after 1994, with provincial government departments forming the middle layer between local municipalities and national government. The national Department of Water Affairs and Forestry[6] does not quite conform to the three-tier pattern, as it has no provincial department, but rather regional or 'satellite' offices within the national sphere.

DWAF regional office and local municipalities

The responsibility for water in the three catchments making up the current Inkomati Water Management Area (Sabie-Sand, Komati, and Crocodile) lay with the DWAF head office until 1997. There was a small branch in the provincial capital, Nelspruit, too, but it was mainly concerned with liaison and monitoring. It eventually became a fully-fledged regional office after 1997, the DWAF regional office (Faysse & Gumbo 2004). On 1 July 2003, local municipalities became responsible for water services delivery, whereas the responsibility for water resources management still rested with the national department and its regional offices. The Inkomati WMA comprises three districts – the Ehlanzeni, Nkangala and Gert Sibande. The Ehlanzeni has five municipalities, of which one is the Nkomazi, roughly overlapping with the Nkomazi portion of the former KaNgwane homeland. This region is also referred to as the 'Lower Komati' as it is in the lower reaches of the Komati River catchment.

Provincial Department of Agriculture and Land Administration

The Department of Agriculture (DoA) and Department of Land Affairs (DLA) under the then Ministry of Agriculture and Land Affairs had their provincial equivalents in the Mpumalanga Provincial Government Department of Agriculture and Land Administration (Mpumalanga DALA, hereafter DALA for the sake of brevity), which was tasked with providing support to all farmers in Mpumalanga. When the administration of the homeland of KaNgwane was disbanded in 1994, many of those who were employed there were absorbed into DALA. The head of Technical Support Services at DALA, an elderly, energetic white male intent on reviving smallholder agriculture, was very much involved in water allocation issues in the Nkomazi area, and used to work in the agricultural department of the former KaNgwane administration. The department also dealt with land reform, and had a restitution and tenure upgrading committee.

The Mpumalanga Co-ordinating Committee for Agricultural Water (MCCAW)
The Co-ordinating Committee was the forum for representatives from DALA and the DWAF regional office to meet and discuss issues related to agricultural water use. DALA chaired the secretariat, and the purpose of the MCCAW was to recommend proposed water use projects, drawing on government resources.

The Catchment Management Agency (CMA)
The Inkomati CMA, the first – and, at the time of research, only (of 19 proposed) – was established in March 2004, after seven years of protracted stakeholder negotiations (Anderson 2005).[7] It was officially launched on 2 November 2006, although at the time a number of positions had yet to be filled. However, the Governing Board, which was a statutory stakeholder platform, had been established and comprised 14 members led by a chief executive officer. The board met at least twice annually, and was accountable to the Minister of Water Affairs and Forestry, whilst the individual board members were expected to report back to their constituencies, although many of the people we spoke to had no idea who their CMA representative was. An advisory committee was put together by the minister, to provide advice on which sectors and groups should be represented on the Governing Board. Each group or institution thus nominated four candidates, whose names were submitted to the minister, who then selected one name for final appointment (Governing Board representative for civil society, interview, 1 August 2006). A core challenge of the CMA, according to its chief executive officer, was to help 'transform the water sector...reallocation can never take place in the present situation. They need to get the process of demarcation [making Irrigation Boards more inclusive and representative through transforming them into Water Users Associations] in place first' (CEO interview, 14 July 2006).

There were also Catchment Management Committees, a further delegation of responsibilities according to geographical or functional divisions. Although the purpose of the CMA was to co-ordinate the interests of existing and potential water users and to represent all interests, in particular those of disadvantaged individuals and sectors, the extent to which this was actually the case was doubtful. Many stakeholders did not have the time, knowledge or resources to participate in a meaningful manner, and many also experienced a degree of 'participation fatigue' after the sundry consultations to which they had been invited (Anderson 2005). There was a lack of a common vision of what the CMA should actually be; whether it should 'simply...provide more funding for water management costs' or serve as a 'vehicle to aid the plight of the poor by providing water licences to emerging commercial farmers' (Anderson 2005: 5). The CMA was supposed to be self-financing, with water users contributing to management costs through the payment of 'water user charges'.

Non-governmental organisations (NGOs)

There were a few NGOs present in the area as well, such as AWARD,[8] which was quite influential and did important work on water supply issues, mainly with rural communities in the Sand catchment area, and GeaSphere,[9] which campaigned to bring attention to the negative impacts of the commercial forestry sector.

Patronage and paternalism

What were the discourses dominating in the Inkomati? As argued in Chapter 3, the ways of speaking, choice of categories and terms used all serve to create particular social identities. At the policy level, this implies that a diversity of existing water uses was collapsed into a 'basket category' of existing lawful uses and the potential 'emerging' users that are targeted in the reform, in other words, the historically disadvantaged individuals. The 'subject positionings' existing in the Lower Komati were more fragmented and complex than those existing at the national level. Rather than users simply being referred to as 'historically disadvantaged individuals' (HDIs) versus 'existing lawful users' (ELUs), the fault-lines ran more along irrigated sugar – comprising both the Nkomazi Irrigation Expansion Programme beneficiaries (see below), as well as the white, commercial farmers – versus dryland farmers and domestic users, with a concomitant struggle for authority and contestations between the Department of Land and Agriculture (DALA) and the DWAF regional office.

Sugar in the Inkomati: A recipe for success?

Sugarcane was still a key crop in the Inkomati WMA. On the eastbound N4 out of Nelspruit towards Malelane and Komatipoort on the Mozambican border, there were scores of farms planted with mixtures of sugarcane, citrus and bananas, with some vegetables and macadamias interspersed. The tributary to the Komati, the Lomati River, served as a natural boundary between territories, with emerging farmers on the right bank and established commercial farmers on the left.

Emerging farmers

At the beginning of the 1990s, the then KaNgwane Administration initiated the Nkomazi Irrigation Expansion Programme (NIEP) to promote small-scale farming by black farmers. The emphasis was squarely on sugarcane, with some additional vegetable projects for subsistence purposes. Sugar was chosen primarily because it was considered to be a very easy crop for small-scale farmers to grow; there was also a guaranteed market providing a dependable income and it could be harvested in the first season (Head of Technical Support Services at DALA, interview, 23 June 2006).

The decision to build the Driekoppies and Maguga dams played an important role in the development of the NIEP scheme (Brown & Woodhouse 2004), although the perception among some of the farmers was that the dams were built primarily to supply the chiefs (chairperson of Cotton Farmers' Association, interview, 29 November 2006). As noted by Nxumalo (1997: 7), the Driekoppies Dam was intended primarily as a supply of irrigation water and came at a time when there was merely a '…patchwork of fragmented, old and poorly maintained water supplies…in the districts. These consist of boreholes and a few river water purification/reticulation plants. Due to population growth and movements, only a handful of about 30 communities have greater than 50 per cent access to safe water.'

In 1994, when the KaNgwane homeland area ceased to exist, as the new democratic government took office and the old regime of the homelands became a thing of the past, many people that used to work for the KaNgwane administration moved into the offices of DALA. From this new vantage point, they continued to administer the NIEP scheme. The scheme came to include some 1 200 black farmers, with 13 projects in the Malelane area that abstracted from the Lomati River, and 22 in the Komati area. On the Lomati, the left bank was a mainly 'white' area, using some 90 per cent of the water, with the right bank bordering on what used to be the former homeland of KaNgwane. The cane is delivered to the sugar mills at Malelane and Komatipoort, both operated by the TSB, and net income is returned to the farmer after deducting water and electricity costs. SASA, the South African Sugar Association, is a partnership of millers' and growers' associations and runs an extension service, SASEX. All sugar farmers in the Komati area, both large and small-scale, are members of the Mpumalanga Cane Growers' Association (MCGA chairperson, interview, 9 June 2006).

The projects were developed in two phases, the first initially aiming to support some 950 farmers on about 7 000 ha, with an average size of 7 ha per individual plot (TSB staff, interview, 8 June 2006). In 2000, a second phase was made possible with grants from the Land Redistribution for Agricultural Development (LRAD)[10] scheme. This second phase became known as 'the seven', and comprised the projects of Phiva, Mzinti, Magudu, Sikhlawane, Nthunda, Langeloop 2 and Vlakbult (SASEX extension officer, interview, 25 May 2006).

The sugar projects were generally viewed favourably in the media and national policy circles:

> The small-scale sugar cane project is one of seven projects in the region aiming at helping previously disadvantaged South Africans, especially

women, to join more than 50 000 already established small sugar cane growers and tap into one of the country's biggest foreign exchange earners.
(African Connexion 2002: 28)

The role of the sugar industry was crucial in helping black farmers to establish themselves on smallholdings and guaranteeing an outlet for their crop. This is a very visible effect. In the context of the disappointingly slow pace of land reform, the small cane grower schemes have been heralded as a success.
(Lorentzen & Cartwright 2006: 18)

At a national policy level, from the point of view of the people participating in the water allocation reform process, the sugar projects in the Inkomati were seen as backing their ideas about promoting 'productive' and 'efficient' use of water for HDIs. The DFID consultant spoke approvingly of them, stating that sugar was promoting small-scale farmers by providing support and ensuring stable prices. According to him, sugar was a successful crop, one should 'put it' in sugar, observing that 'you see them [the emerging farmers] driving around in their little *bakkies*,[11] and there is the potential for local economic benefits to accrue.' He saw sugarcane in the Inkomati as a way to 'straddle groups one and two,' meaning the 'first and second' economies, through building up black farmers as commercial sugar growers.

FIGURE 4.3 *CMA members in discussion at the Maguga Dam*

FIGURE 4.4 *Ronnie Morris and Mr Brown, a commercial farmer*

FIGURE 4.5 *Working in the sugarcane fields, Mzinti, Nkomazi*

FIGURE 4.6 *Sugar fields stretching to the horizon, Madadeni, Nkomazi*

FIGURE 4.7 *Mr Nkalanga in his field at Spoons 7B, Komati River*

FIGURE 4.8 *The Khutselani women's group growing vegetables, near Driekoppies*

FIGURE 4.9 *Mr Edward Sambo in the pump house at Masibekela*

FIGURE 4.10 *Masibekela farmers' association next to the dam*

However, the positive image of the sugar schemes in the media contrasted with local opinions of what was actually happening to the schemes. One of the extension officers working in the area, Banie Swart, wrote an evaluation report on the projects. In the report, he noted how the sugar prices were good and the energy prices were low, back when they started up. Most of the farmers were able to make enough money to raise their standard of living and boost the economy (Swart 2006). But over the years, returns from ratooning[12] diminished, yields declined, soils were depleted, and new projects were developed on marginal soils. Input costs increased and the sugar price stagnated and even declined in real terms (Swart 2006: 2). In Swart's own words:

> Some sold their land; others simply abandoned their fields; some started as labourers or started up informal businesses. The yields came down, and the farmers are becoming poorer. They don't have enough money to put petrol in their car. If you produce 75 tonnes, you would be able to service your loans, your furniture account, your food account, but you would have nothing to put in your pocket. You do not contribute at all to the local economy... Basically, it's the law of diminishing returns, and the yield for each project is going down. Now the yields are down to 44 tonnes per hectare. Financial

rewards from sugarcane are currently very moderate and require exceptional management skills. With current world trend forecasts, many farmers will not survive the next two years.
(Interview, 25 May 2006)

FIGURE 4.11 *Graphic representation of yield decline, Komati*

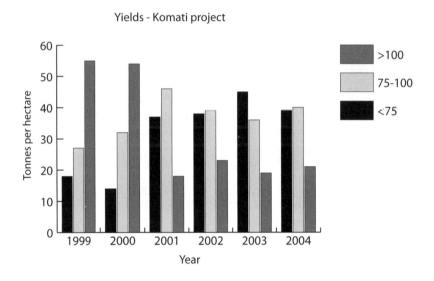

FIGURE 4.12 *Graphic representation of yield decline, Lomati*

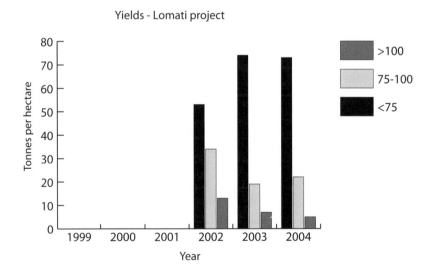

A second round of projects, 'the seven' (see page 94), started in 2003. The larger portion of the latter was developed on soils that, in Swart's opinion, should never have been cultivated at all. These soils were of low potential, with low water-holding capacity, and often with severe drainage problems. Warnings from extension officers and soil scientists were ignored. Community pressure to go ahead was a major factor, and the comment was made that soil problems could be rectified later. The extension services of SASRI argued that inherently poor soils could not become good soils. The SASRI head of extension wrote to the MEC of Agriculture, expressing his concern, since it would be the extension services that were ultimately held responsible. After only three seasons, the cane yields from some of these projects were disastrous.

Apart from inherently poor soils, there were also salinity problems. The cane looked as though it had been affected by drought, but this was more likely sodium toxicity, due to the tendency of small-scale farmers to over-irrigate. The emerging farmers were less restricted, and effectively received 20 per cent more water during periods of drought than the commercial farmers, who may have had some on-farm storage. All the small-scale farming projects were upstream, so they should in theory have had an advantage, said Swart. 'But they don't use the system, so they don't have enough water. What they do, is to pump as much as they can,' he stated, continuing, 'they practice "the wetter, the better", irrigating over the weekend, leaving their sprinklers on for 24 hours a day, even when it has just rained' (interview, 25 May 2006).

An illustration of this was given by Mr Mhlongo, the grandfather of one of the students I had got to know at the Lowveld Agricultural College. He was a sugar farmer in one of the older schemes, the Langeloop 1 (Mabodweni) on the Lomati River. He related how this particular project had been established in 1996, and that there were currently 55 members, each farming seven hectares. In his view, the decline of the sugarcane fields was mainly to blame on their lack of storage; 'we are no longer able to irrigate 24 hours a day.'

The chairperson of the Komati River Irrigation Board, himself an emerging farmer, explained how

> the sprinkler irrigation systems are wasting water; there is a lot of evaporation, drip irrigation is much better because it conveys water directly to the plants, and allows water to get into the soil. But it is very expensive to implement…it is possible to get grants and loans from the Land Bank. But the problem is that in the first year, the yields are OK, then the second, they fall a little, then the third, they are average, then the fourth, and they are low. At the same time, you have a loan with interest rates that you need to pay back, at the same time as your output is becoming lower. That is the problem. Also, you have to pay on a monthly basis…that is difficult, because your income is not necessarily on a monthly basis.
> (KRIB chairperson, interview, 7 June 2006)

Land access

A contributing factor to the falling levels of productivity was, according to the extension officer, the system of land access. As noted earlier, during the apartheid era, chiefs often became puppets of the regime, which built on the 'indirect rule' policy of the British (Mamdani 1996). Chiefs amenable to the government line were placed in charge of tribal areas, where they acted as 'clenched fists', with few of the traditional trappings that used to function as checks and balances to chiefly power, such as councils (Levin & Mkhabela 1997). This situation led to the chiefs having little credibility in some areas, and traditional leaders were often treated with disdain, in part due to their links with the apartheid government. After the transition, the ANC vacillated with respect to the role of the chiefs, and coalitions were formed with what were regarded as 'progressive chiefs' in several homelands, amongst them the KaNgwane (Levin & Mkhabela 1997). However, land access in many areas still depended to a large degree on chiefly authority (Ntsebeza 2000; Ashley, Chaumba, et al. 2003; Lahiff 2003; Cliffe 2004; Claassens 2005).

In the Lower Komati, perceptions of chiefly power varied widely amongst the whites. Few of the college employees, government officials (apart from the Head of Technical Support Services of DALA who had been part of the KaNgwane administration), white extension workers or TSB employees knew much about the chiefs in terms of how many there were, what territories they had under their jurisdictions, and what kind of authority they exercised. Some people, especially white farmers, disregarded the notion that the chiefs exercised any authority to speak of. As a chairman of one of the Irrigation Boards expressed it, 'the chiefs are a toothless bunch, as most of their powers have been taken away... They are involved in all the meetings, but they don't have much power' (Irrigation Board chairperson, interview, 23 May 2006). Later, however, we toured the project area with one of the Lowveld Agricultural College senior lecturers, who observed that 'once you get off the tarred roads, it's the chiefs who rule' (casual conversation with agricultural college lecturer, 23 May 2006).

In the Nkomazi area, there were seven chiefs, representing a corresponding number of tribal authorities. There were four tribal authorities in the Nkomazi East area (east of the Komati River): Lugedlane, Hoyi, Siboshwa and Mhlambo; whereas Nkomazi West was made up of three tribal authorities: Mawewe, Mhlaba and the Matsamo – which included some 14 out of the total of 52 villages in the whole area (both east and west), and thus was quite large. Several tribal authorities, including the Lugedlane and the Matsamo Tribal Authority, were chief-less at the time of research – the chief at Matsamo, Shongwe, had died about six months earlier. I was told by a female farmer in the Figtree scheme at the Hoyi Tribal Authority that the succession struggle to replace the deceased Matsamo chief was still going on, as there were several contenders – 15 in all – for the position of chief. In the meantime, his widow was acting as head.

The chiefs in the area still wielded a great deal of power in terms of land allocation, through the issuing of Permission to Occupy certificates (PTOs),[13] although there were motions to upgrade tenure under the land reform programme. Although the chiefs largely retained the power to reallocate land once it was allocated by means of a PTO, the 'present land-tenure system offers relative security' (Commission on Restitution of Land Rights 2007: see also Levin & Mkhabela 1997).

However, when talking to people in the area, the impression was more nuanced. Rather than a conventional PTO, sometimes chiefs would issue lease agreements. In one case, a farmer's association established for the purpose of producing cassava had obtained a lease agreement in 2002 to occupy a 234.68 [sic] hectare plot for '9 years and 11 months' from the chief of the Matsamo Tribal Authority (chairperson Siyavuma Cassava Farmers' Association, interview, 19 October 2006). No production was taking place at the time, allegedly due to the lack of water, and the chairperson of the association explained that the lease agreement would only be renewed if the association were granted a water-use right, as this would increase the chances of the land becoming productive. Thus, in this case obtaining a water-use right was a means of strengthening a tenuous claim on land. Though water-use rights were not accepted as collateral by the banks (Land Bank employee, interview, 5 September 2006), the fact that water rights were issued by the government represented a form of security, in that having a water right increased the productive potential of the land.

In some areas, such as that under the late chief Lameck Mbhunu Ngomane II of the Lugedlane Tribal Authority, people were expected to pay R600 for a PTO – levies were charged per person rather than per hectare (interview, Steenbok farmers, 20 June 2006). The chief was, therefore, trying to get as many people involved in a project as possible in order to increase his revenues. For example, the Mfumfane B scheme (one of the phase I schemes) did not yet have water. We searched out the secretary of this scheme – a widow and mother of two in her early thirties. When we met she explained that initially there had been a number of people involved, each with a 10-hectare plot, but then the chief had decided that these plots should be subdivided. Most people had to split their plots in half, although some kept seven hectares:

> We want to plant sugarcane, but we are left behind. So we are planting with cotton...we felt abused by the chief when he said that we should split our plots into two five-hectare ones. We had to give the plots to the chief, and then he would decide who was to get them, and the new owners would have to pay the chief to get the land. We did not get anything in return for giving up that land. (Mfumfane B secretary, interview, 8 September 2006)

With respect to the new developments – 'the seven' as they became known – there were several reports of trouble. All of these developments, with the exception of one (the Magudu project in Mawewe Tribal Authority), were implemented within the boundaries of the Matsamo Tribal Authority. In the Vlakbult scheme in particular, there was infighting reported by the extension officer, and by farmers. The reason, it appeared, was that the late chief Shongwe of the Matsamo Tribal Authority had decided to increase the number of beneficiaries on the scheme. One farmer from the area explained that 'two groups are fighting; the chief divided members fifty-fifty, and now one group is disputing that, because the other groups are newcomers' (interview, 19 October 2006). On the same occasion, another farmer in one of the already-established schemes (the Langeloop) under the same tribal authority, attested to the prevailing sense of insecurity: 'there is no security of the land, but we have been negotiating with the Department of Land Affairs to negotiate with the tribal authority to get a title deed, but nothing has come of it yet.'

The tenure upgrading project in the Ehlanzeni district aimed 'to assist in the tenure upgrading within existing settlements within the next 10 years, and assist in the formalisation of tenure for labour tenants' according to the young, black female director of the restitution and tenure upgrading project for the Ehlanzeni district (which included the Nkomazi municipality) at DALA. She explained that residents in R188 areas[14] would be allowed to choose the kind of tenure system they preferred under the upgrading scheme. With respect to the Communal Land Rights Act (see, for example, Claassens & Cousins 2008), she explained that

> the CLaRA addresses the issue of people converting a PTO to title deed...If a person has been using the land for many years, and he wants to own it, then they send that particular person back to the community to decide. In most cases, the community does not agree.
> (Interview, 22 August 2006)

The question, then, was who was the 'community' in this regard? She explained that even though there was little change in existing patterns of land tenure – the CLaRA had not yet been implemented – the land reform would provide new opportunities for communities to choose whether they would like to have title deeds or to retain the PTO system. An important case was the huge 30 000 ha Tenbosch area at the south-eastern end of the Kruger Park. Much of this land was farmed as sugarcane estates by the TSB, but was under claim from a total of 15 000 households from the Lugedlane, Siboshwa and Hoyi Tribal Authorities, who had been removed from the area between 1953 and 1970. Many of the farmers we talked to at Lugedlane and Hoyi were party to these claims, and wanted the land to be held by title deed (PLAAS 2004).

Though the security of the PTO was tenuous, chiefly allocation in the first place tended to favour close kin and relatives. An example was the meeting with Mr Shongwe at the Matsamo Tribal Authority offices in Schoemansdal. In his early forties, he was a member of one of the sugarcane schemes in Matsamo, Sikhlawane. He had also been the deceased chief's driver. After having spent the better part of the day walking around the project area and learning how it was run, we broached the subject of land allocations and how people got access to the projects. Shongwe smilingly responded, 'I'm in the chief's family, so it was easy...I just went to him and explained that I wanted to be part of the project, and then I was provided a PTO' (interview, 27 September 2006).

He explained that, initially, all members of the project had been given a similar sized plot, but then the chief had allowed some of the members to obtain part or the whole of other people's plots, so that now members' holdings were highly irregular. Some had five hectares, some had 10, and others had 15. This pattern generally seemed to be the case with the older schemes. As the project evolved, there was increasing flexibility in terms of who was actually holding how much land. In the newer projects, all members had equal holdings, usually about seven hectares in size. Chiefs influenced who became beneficiaries of development projects. For instance, although the LRAD scheme was intended to distribute land to previously disenfranchised households, the participation of the tribal authority in determining ownership of the plots had important implications for the effectiveness of these projects in reaching the poorest members of the community (King 2005). Of the 35 plots demarcated for the Mzinti community under the Matsamo Tribal Authority, the majority belonged to the households within the older section of the community, who had strong ties to the chief (King 2005). In a number of cases, the farmers pooled together as collectives and were effective in lobbying the tribal authority for land. In others, the tribal authority used the project to recognise existing power structures within the community, rather than alleviate poverty by granting occupancy rights to its poorest members. The political elite benefited from the distribution of plots, which suggests that tribal authorities continue to use development projects to control resource and reward allies within KaNgwane (King 2005).

Gender relations

Although the sugar schemes had been advertised as promoting the role of women as well (see the quote from *African Connexion* above, which states that the small-scale sugar schemes were aimed at helping 'previously disadvantaged South Africans, especially women') the picture was more complex. A Development Bank of South Africa (DBSA)-funded survey in 1998 reported that although one quarter of the NIEP sugar projects it funded were for women, this did not imply that they were in control of the land. Rather, the report argued that many of them were labourers on land controlled by male kin. Though women were ostensibly managers of their individual plots, it found that

> most women registered as sugar cane farmers as part of the Nkomazi Irrigation Expansion Programme do not, in reality, control their farms as legitimate decision-makers...Farms registered in the names of women are generally controlled by their sons, families of their deceased husbands, working husbands or aged fathers.
> (*Mail & Guardian* 11 November 1998)

Where women did have land of their own, this was often as a result of the death of the original holder. During a group meeting with farmers from the Figtree schemes in the Hoyi Tribal Authority, my question of how many female farmers there were elicited the following response from a male farmer of the Figtree B scheme: 'There are many women farmers here...because all the men are dying' (interview, 20 June 2006). One such female farmer was Dudu Lubisi of the Figtree A scheme. In her early thirties, she told us how she had inherited the plot from her grandfather and was now working it on her own in order to support her mother and her two small daughters. She explained that 'there are more women like me in the schemes', but that they had to share the land with their brothers or other members of their father's extended family (interview, 26 June 2006).

Although women held plots individually within the schemes, there were no female chairpersons of any of the schemes we visited. All the chairpersons were male, with women sometimes holding positions as secretaries or clerks (for example the Ngogolo, the Lugedlane, and the Magudu schemes had female secretaries and clerks). Women's groups had been allotted minor portions of land in several projects. Such groups were established as a sewing group or women's club, and some then lobbied the chief to gain access to land in the sugar scheme. There was a women's club portion in one of 'the seven', in Magudu in the Mawewe Tribal Authority. There were at the time 106 members in the scheme, which totalled 468 ha, figures that departed from what was provided in the 2005 extension report. Of these, 46 were women, who were organised as a women's club and held 23.6 ha in total (interview, women's club member at Magudu, 29 November 2006). Passing that particular plot, it was evident that it was one of the best tended, a fact readily acknowledged by the chairperson of the scheme.

These observations of how chiefs were exercising their authority in providing land for those close to them, and of women mostly accessing land through the death of their male kin, are reinforced by other authors such as King (2005) and Brown and Woodhouse (2004). They also concur with Rangan and Gilmartin's (2002) observations, from the area around Malelane, that the Constitution's commitment to gender rights is incompatible with the formal recognition afforded to unelected traditional authorities, a contradiction particularly evident in land reform where women's right of access is denied through the practice of customary rights.

Water control and capacity constraints

Whilst the white, established farmers had individual pumps, the farmers on the sugar schemes shared one or two pumps and took turns in watering their plots. There were no gravity schemes. Water was pumped into a balancing dam or directly onto the fields, and distributed to the individual blocks via rotation. Most schemes used dragline sprinklers, which could be rotated on the plots. Whereas the commercial farmers often lived on their farms, the small-scale farmers' schemes were usually located some distance away from the villages in which they lived. Although a few schemes had balancing dams, none had storage dams, the lack of which became a recurring issue during our visits to the projects. An example was the Figtree B project, established as a cotton project in 1985, and then converted to sugar under the NIEP in 1991. The chairperson, a middle-aged emerging farmer, told us that if they could have storage dams it would be acceptable to have meters to monitor use rates:

> We have water rights in hectares, and when there is too much water, we can pump as much as we like...farmers have to be educated about irrigation... when we are pumping 24 hours a day, we are wasting water. The water needs to be controlled, by having meters and storage dams, so that we can share with the ones who do not have water at the moment, like the cotton and maize farmers...but the problem is that the government was providing the white farmers with subsidies for their storage dams, but they have not provided us with the same kind of support.
> (Interview, 22 June 2006)

There was also a severe shortage of capacity in extension services to deal with farmers' problems. Philemon Mthembu worked as a control technician at one of DALA's district offices and was responsible for the whole Nkomazi East area. In his view, the later projects 'should never have been developed'. He related how 'somebody came along from the Department of Agriculture, meetings were held with the tribal authority, and everything was done in a hurry' (interview, 1 August 2006). Echoing the extension officer quoted earlier, he said that several projects had been started on marginal lands with poor soils that should never have seen sugarcane, but that the farmers settled on those projects did not want to stop farming. One farmer's explanation for wanting to stay put, was: 'I'm 55 years old, I cannot start to plant trees, because by the time they will be producing something, I will be dead.' Mthembu observed that there were only a few young farmers on the sugar projects, the majority being mature in years. He added, 'since it's considered such an easy crop to grow, people don't really stay around much'.

In terms of financial management capacity, Mthembu remarked that 'they [the small-scale farmers] are squandering money; they don't know how to manage their finances. When they get paid, they go and buy smart clothes, they go and buy a *bakkie*, but they don't invest, they don't think about the future.' He explained how farmers went away to Johannesburg and left their plots for weeks on end, and how many of them worked as teachers and clerks in addition to farming their sugar plots. Umthombo Agricultural Finance took care of farmers' finances. All costs, including electricity, were subtracted when the sugar was delivered to the mill, so the farmers got the majority of the balance. The secretary of the Mfumfane B scheme, Olga Mavundlu, told us how one could tell that the sugar payments had arrived: 'They [the men] go drinking' (interview, 1 August 2006).

Hence, although the NIEP scheme was presented as a potential success in terms of getting emerging farmers on the 'path to commercialisation', it was struggling and facing problems with maintaining productivity, owing in part to land access dynamics, as well as a lack of capacity and control on the part of the farmers involved. Moreover, a report commissioned by the World Bank in 2006 and funded by the World Wildlife Fund and the Dutch Government highlighted the possible negative impacts of liberalising the sugar sector on poverty alleviation efforts, in particular emphasising trade-offs between liberalisation and environmental integrity (Lorentzen & Cartwright 2006).

Established farmers

The white farmers in the area have enjoyed the strong organisational support of Agri-SA, historically a white farmers' union. Estimates of the number of white, commercial farmers in the Lower Komati area varied. A representative from the TSB reckoned that there were some 120 all in all, farming about 20 000 ha, of which about 80 to 90 were drawing water from the Komati River, and another 30 to 40 farmers depended on the Lomati (interview, 8 June 2006). Twenty-six of the farmers along the Lomati River had been involved in a dispute with DWAF over water rights (senior adviser, water resources and planning, interview, 25 April 2006). Apart from the general support of AgriSA, the sugar-growing farmers also had a firm support system in the sugar sector. Moreover, most commercial sugar farmers also tended to grow sugar in combination with other crops, such as citrus, bananas, or vegetables, thereby effectively spreading potential risk. Most were middle-aged or elderly white males. On one occasion we met a young woman who was running the farm on behalf of her father, who was away in Mozambique on business.

Traditionally, commercial farmers have organised water abstraction matters through Irrigation Boards tasked with controlling their riparian river water. The new legislation declared that these boards are to be transformed into Water Users Associations, but transformation has been painfully slow to materialise.[15] In 1957, the Komati was declared a Government Water Control Area (GWCA), which meant that the farmers' riparian rights were attenuated through the imposition of quotas or permits. According to the 1956 Act, permits could be classified as normal flow permits and surplus flow permits. A normal flow permit indicated that one could take a 'reasonable' portion of the allocable water. Permits for surplus flows allowed farmers to store flows in excess of 'normal' flows. The size of quotas in the Komati and Lomati rivers were determined in the 1970s by an engineer, Charles Sellick, who calculated that farmers on the Komati River could pump 9 950 m^3 per hectare per year. The permits for the people on the Lomati River were set lower, at 8 500 m^3 per hectare per year, owing, Sellick argued, to the difference in the climatic conditions of the two rivers and the increased rainfall falling on the Lomati side. Where the permits on the Komati River were stipulated in a government notice, the Lomati farmers' riparian rights were not formally regulated by any permit, but the 8 500 m^3 has been taken as a convention (Brown & Woodhouse 2004). Farmers would pump water according to their quota, and when there were surplus flows, they would pump this into storage dams on their farms.

The 'water-per-hectare' terminology still lingered, though the new Act stipulated that water rights should be registered in terms of volume (cubic metres) only, rather than as volume per hectare as in the quotas. When asked about their water rights, most farmers would respond in terms of 'I've got 250 hectares of water rights from the Komati,' or 'I've got 122 hectares of water rights from the Lomati.' The staff at the regional office acknowledged that 'farmers still tend to think in terms of water rights as hectares' (interview, 4 August 2006). Several of the farmers interviewed felt that licences offered little incentive to use water efficiently, 'since if you don't use it, you lose it' and therefore if they converted to more efficient irrigation systems, they would rather expand the area under cultivation than return the unused flows. This view was corroborated by the SASEX extension officer, who argued that 'the thing is, what is happening now, is that you pay for a water allocation, and therefore you use the entire allocation, no matter whether you need it or not. You store it in your soil but without regard for the crop requirements...there is no incentive to save water. So what many farmers do is to extend their land under cultivation, to make full use of the water. Especially the emerging farmers, they irrigate over the weekend, and leave their sprinklers on for 24 hours. They don't have any incentive to save water at all. Why should they?' (Interview, 25 May 2006).

Both Irrigation Boards had installed a Water Administration and Measurement System (WAMS) to deal with the increased problem of theft. At the time of research, the Lomati River Irrigation Board employed a youngish engineer whose responsibility it was to co-ordinate the Irrigation Boards and KOBWA in terms of water distribution along the Komati and Lomati rivers. He explained how the system measured how much of each farmer's quota was used by monitoring their pumps, and automatically shutting off flow when the quota was reached. The floods in 2000 washed the system on the Lomati River away, after which it was not been re-installed on that stretch (interview, 24 May 2006). The WAMS system effectively releases farmers from having to cooperate with one another, and water theft under this new system, according to one commercial farmer quoted in Waalewijn (2002: 55), feels more like 'stealing from the government.'

The distribution of water amongst farmers was relatively complex, since each had individual requirements depending on the types of crops, cropping patterns, and nature of soils on the farm. In the opinion of a senior lecturer at the Lowveld Agricultural College, most farmers do not really schedule, in other words, they do not plan according to crop requirements in a 'scientific' fashion. 'There are a few hi-tech types that use tension meters and neutron probes to gauge moisture content, but most follow a rule-of-thumb approach' (interview, 23 May 2006).

All the commercial farmers had individual pumps on the rivers, and most had storage dams. Although the storage dams were often filled up with water from the river, which would be part of the farmer's water-use right, they were, in effect, a way to secure surplus water such as catchment flows, flood water, small streams, rainwater, and storm water. The impoundment of the latter had not been recognised by the 1998 Water Act (see Chapter 2: 36). An illustration of this was given on a visit to one of the large estates bordering on the Lomati River, which used advanced technology to measure soil moisture content and to monitor its water usage of 1 100 ha water rights from the Lomati River, in addition to that from boreholes. The farm's owner, Jamie Chance, a middle-aged, forthcoming man, offered his opinions freely on the prevailing practice of water allocation.

> We've got a total dam storage of 850 000 m³...we can pump into our dams; plus we got basically free water coming down from the Sterkspruit River... That is a real plus; we got dams...a lot of other farmers that draw water don't. In times of drought, one goes on to restrictions, obviously. We, meanwhile, have got dams on our farm. That is a major, major thing that should happen; there should be many more storage dams...one should get away from licences, but just pay for the water one actually consumes.
> (Interview, 6 September 2006)

We had heard much the same story from other commercial farmers. One farmer on the Komati side, Mr Terence (he wouldn't give his last name) argued that in the 'old' days, when there was no sugarcane, the quota of 9 950 ha was sufficient, but sugar needed way more than that – it needed about 13 000 ha, in his opinion. But now they no longer have the 9B quota as stipulated by the 1956 Act, the 'surplus flow' quota (see Chapter 3), so he has to rely on his dams, arguing that 'it's part of our cultural practices [to build dams]' (interview, 3 August 2006). In a sense, then, the notion of filling up dams with 'surplus flows' is still subscribed to by some of the commercial farmers.

Generally, the impression was that the established commercial farmers were characterised by strong alliances which were resistant to change, and were still very much caught up in the terminologies and practices of the old 1956 Water Act. They had a good deal of flexibility in that they did not depend solely on sugar for their income, but diversified into bananas, citrus and other crops as well. In addition, they enjoyed a high degree of 'hydrological security' in that they were able to store river water, in addition to catchment flows and storm water, for use during the drier months when irrigation was restricted.

Perceptions of sharing and co-operation

As has been described, a number of the emerging sugar schemes were struggling – particularly the more recent schemes. How did this influence the relations between the commercial and 'emerging' sugarcane farmers?

Many of the commercial sugar farmers we spoke with exuded a strong, racially tinted, paternalism. For example, during a meeting with the chairperson of the Lomati Irrigation Board, my research assistant had to leave the room due to his increasing embarrassment at the chairman's expositions about racial realities in the region. The chairman, who had been an 'unwitting member of the apartheid government', ventured to explain the social fabric of the Inkomati. He shared his views on how the area used to be a 'harmonious' place:

> Everybody lived in their kraal and wore traditional clothes, and you had grandma and everyone living as extended families. Now there's been a vast change in this area during the last forty years; it's no longer like it used to be. They lived off the land, but now there is this very Western culture, where everyone has his own cell phone. The family system was lost somewhere along the way. We have problems with theft; people steal all kinds of things, copper wire, cables. It's not like in the past.
> (Interview, 24 May 2006)

He told us how he had tried to set up an advisory group of 30 or so people to discuss farming-related problems at his own farm, and related how there would be five to seven different factions within the group, who did not trust one another. 'It wasn't a tribal thing, but mainly stemming from the chequered past that South Africa has experienced.' He explained the tensions as having been caused in part by the mass migration of people looking for work in the mines and leaving for big cities like Johannesburg to try to find jobs. In his view, this weakened family ties and had created social disruption: 'Husbands find new girlfriends; you know, that sort of thing.' He regarded the sugar schemes as wasteful, but necessary for the maintenance of social stability, a view that resonated with those of other farmers in the area. There was thus a keen sense that the sugar schemes contributed to social stability in the region. In the words of the SASEX extension officer: 'We need to hold this thing [the sugar projects] together. If this comes apart, everything comes apart' (interview, 25 May 2006).

But whereas there seemed to be a prevalent feeling that emerging farmers should have water, the commercial farmers did not see the logic in curtailing their own water use in order to share it. Reactions varied from the mildly resistant to more irascible outbursts:

> Nobody has a problem with the fact that a certain group is entitled to water, but it is the way that it is done that is going to be problematic. If I was to be asked to give up 20 per cent of my water to someone, I would have to be convinced about the equity benefits of that.
> (Lomati Irrigation Board chairperson, interview, 24 May 2006)

> I think it's unfair to take something away from somebody. I think it is unfair to take productive water away, where you can't guarantee that it is going to be productive. I think they must buy farms and properties and do the reallocation by land. Also, the infrastructure, everything; I think it's not productive to take [it] away. If there's water that's not been used, lying around, then reallocate it, but water that's being used; it's not fair.
> (Commercial farmer on the Komati River, interview, 4 August 2006)

> We are not farming crops, but managing water, converting water into cash... If you take away some of my water, and then ask me if my business is still OK, it is like cutting off my fingers and toes; but yes, I'm still walking.
> (Commercial farmer on the Komati River, interview, 4 August 2006)

Thus, although there was a prevalent paternalistic attitude, with the notion that there was a need to 'hold this thing together', there was little willingness to sacrifice any of their own benefits for the sake of social stability.

Emerging farmers' lack of leverage

The emerging farmers did not feel that they had much leverage in decision-making, although they were nominally members of the Irrigation Boards on a par with the established farmers, and the chairperson of the KRIB was black. During one of the group meetings with emerging farmers, a farmer from the Figtree scheme complained that 'When we come to the Irrigation Board meetings...the agenda is already set...we don't really know what the commercial farmers are doing' (interview, 10 July 2006). Another farmer from Langeloop said that 'they [the commercial farmers] are hiding things...[the] main problem is that some emerging farmers are not educated, and the commercial farmers and the Irrigation Board are taking advantage of this. We lack a proper person who can stand [for us] in the Irrigation Board' (interview, 10 July 2006). I was told that during the preceding year's drought, there had been a number of conflicts between the emerging and commercial farmers, to the extent that they stopped greeting one another: 'We didn't say hello...we thought to ourselves, "He's taking my water!"' (interview, emerging farmer, 20 July 2006).

The emerging farmers got the impression that the commercial farmers were irrigating all the time: 'Their cane is green, whilst ours is dying...a question we are having is, before the dams were full, the commercial farmers' dams would also run dry, but now, after the completion of the dams, they are never dry. Why is this so?' (emerging farmer from Figtree scheme, interview, 11 July 2006). The issues revolved around meters and storage. This was confirmed during an interview with one of the commercial farmers on the Lomati River:

> There was a dispute in the drought last year...They [the emerging farmers] couldn't understand how we were irrigating. We could irrigate 24 hours if we wanted to; they could only irrigate according to the schedule given by the Irrigation Board.
> (Commercial farmer Komati River, interview, 4 September 2006)

In the eyes of the extension officer, however, the problem was not so much one of water as management. The emerging farmers were seen largely by both the SASEX extension officer and the commercial farmers as in need of external management. Relating how 'they [the emerging farmers in the new schemes] were ripped off last year by an independent consultant, who installed an irrigation system totally incompatible with the scheme...we had to go in and fix [it]' (interview, 25 May 2006), the extension officer argued that the solution was to partially outsource management. In his view, letting a corporate body take care of irrigation management would render the projects more sustainable, since the farmers would then not have to make the day-to-day decisions on irrigation, but could let this be handled by outside agents.

Thus, there emerged an alliance of 'patronage and paternalism' between the commercial farmers and extension officers involved in sugar and the emerging farmers, who were part of, or aspired to be part of, the sugar schemes established in the former homelands. Whilst there was a prevailing belief amongst the commercial farmers and the extension officers of the need to 'hold' the sugarcane projects together, there was little willingness to share their own water to this end; getting water to these schemes was regarded as the government's responsibility.

Contested allocations

These alliances were to some extent mirrored in the conflicts over water allocations. There was a prevalent feeling among many of the farmers that they were being short-changed by the upstream water users, who were, in the farmers' minds, the ones causing water scarcity. Other existing and potential users, however, viewed the sugar farmers themselves as culpable. Reasons for scarcity, then, were contested. So was the authority to allocate water at the provincial department level, in part resulting from the fact that an old treaty, dating from before the apartheid state was dismantled, was still adhered to, which created friction between the Department of Agriculture and Land Affairs and the regional office of DWAF. The potential for reallocating water through trading, both intra- and inter-sectoral, was also fraught with its own problems.

Scarcity and the blame game: Sugar versus non-sugar

The notion was widely held that the catchments that had been designated as pilot catchments by DWAF were water stressed, but the senior adviser (water resources planning) within DWAF, who dealt with broad-scale water planning in the eastern region, had this to say when asked whether the Inkomati was stressed:

> Not if you look at the rest of the country; there are very few catchments that are not water stressed. If you look at the eastern part of the country, that's where they have the highest rainfall. You have particular phases of high development levels. You cannot generally say that the eastern part of the country is really water stressed. There is water stress in the rest of the country.
> (Interview 18 July 2006)

This statement provides more nuance than does the Water Allocation Reform policy. It highlights the regional differences, and disrupts the environmental generalisation

(Forsyth 2003) that water scarcity is a ubiquitous and permanent feature. It further underscores the point that scarcity, rather than being treated as a physical fact, should be regarded as a phenomenon arising from the interaction between social and physical systems (see Chapter 3). It is the formation and definition of society's needs versus means that create the perceptions of scarcity; it is not an objective, definable fact outside the social sphere (Aguilera-Klink, Pérez-Moriana & Sánchez-García 2000; Mehta 2005; Mehta 2010). This, then, points to the need to examine more closely the social processes that give rise to the perceptions of scarcity. Such processes are not examined at all in the policy texts; rather, scarcity is taken as a given.

The reasons for water scarcity in the Inkomati are deeply contested, according to Brown and Woodhouse (2004). This view was corroborated by my own observations at the time of research. Broaching the issue with the legal practitioner who worked to assist farmers in the area with their legal dealings with the State in terms of water rights, she noted:

> They said that 100 per cent of the Komati is allocated, but I think that is due in part to lack of knowledge. You didn't see anyone who died of thirst, no children dying of cholera...They are saying that we all got too much, but the land was irrigated, and it is working.
> (Legal practitioner, Komatipoort, interview, 7 June 2006)

There was a tendency amongst sugar farmers to lay the blame for scarcity on other upstream users, in particular Eskom, the forestry sector and Swaziland. In an interview with a commercial farmer on the Komati River, he remarked:

> Right at the beginning, there is Eskom, extracting a lot of water. Next door to them are the big corporations, the Sappis and the Mondis, the forestry...and then it's Swaziland...and then it's us, right at the end...Right at the end.
> (Interview, 3 August 2006)

Eskom in particular was viewed as a big user, and the commercial farmer quoted above went on to observe that 'I mean, I think there's lots of other ways to generate power, than by burning coal...I mean, coal is polluting the area and the water, and it's also consuming lots of water.' I was later told by a senior member of staff at DALA that the commercial farmers had tried to negotiate with Eskom to get them to stop transferring water out of the Inkomati River. According to him, the reason why Eskom was taking water out of the Inkomati rather than out of the Olifants, where the power stations were located, was that the quality of the water in the Olifants was so bad that it could not be used. A group of commercial farmers had reportedly got together to negotiate with Eskom, and offered to pay them a million rand to help clean up the water in the Olifants, so they could cease the inter-basin transfers, but nothing came of it.

Forestry was also under fire from farmers, as well as from a local NGO, GeaSphere. According to Philip Owen of GeaSphere, plantation forestry should be reclassified as agriculture because it is industrial mono-cropping. He argued that

> in times of crisis, when it's incredibly dry, you can stop irrigating. But the timber, it just stays there, and sucks it up...They charge something like ten rand a hectare, which is peanuts for the amount of water they are actually using; the amount of water that they are using is actually staggering. (Interview, 25 September 2006)

Scarcity was seen not so much as a physical condition, but as the result of competing interests, with the commercial farmers regarding the upstream users as the main reason for their lack of water, but hardly mentioning the demands of the upstream commercial and emerging farmers. Scarcity was cast as a result of competing sectoral interests, with the farmers being the losers, since they were at the tail-end of the system, and thus they did not implicate themselves. Rather than merely taking for granted the widely accepted notion of water scarcity as an unquestionable physical fact, viewing scarcity as also a social construct aids in understanding the phenomenon.

Domestic uses

Whilst the farmers viewed themselves as losing out in terms of water access, urban and domestic demand was rising in the area. Domestic water use in the Inkomati was catered for by a mixture of local government, private provision and corporate service operation. Whereas the British corporation Biwater provided services in Nelspruit, it was primarily the local municipality that was responsible for water services provision in the Lower Komati area. According to the chief engineer at the municipality, they were experiencing many problems with regard to service provision, including the theft of water that was used to irrigate sugarcane by scheme members (interview, 26 September 2006). New settlements and demands were increasing at a rate that made it hard to keep up. 'Week by week there'll be new shacks shooting up, new squatters, and they all need services...[the main reason is] migration from Mozambique' (interview, 8 September 2006).

A feature that we came to notice when travelling around the villages in the region was the almost ubiquitous presence of large 5 000 litre water drums in people's backyards. When services did not suffice, one resident explained, the government would send out water trucks. Said a female farmer from the Mfumfane scheme, pointing to her green tank: 'It's political water...you see, when people campaign for local elections, we get water from the local government tankers, but after the elections are over, we do not see them at all' (interview, 8 September 2006). There were also 'independent operators' who would fill drums with water from the Komati River and then drive around the villages selling them. These would cost anything from R100 up to R150 per drum, 'depending on the mood of the driver', as a female resident in Uthokozani put it (interview, 23 August 2006). In a discussion about water use at home, the CMA representative for domestic users told us how many people, women in particular, resented the sugarcane farmers' water use. 'You see, the sugar farmers, both the commercial and the emerging, they are taking all the water from us...It's a problem' (interview, 23 August 2006).

The local municipality had approached the regional office to ask for their help in augmenting water supply. In the words of the deputy director of the regional office, 'the municipalities are very reluctant to buy water; they want us to give them the water. But we can't just take water from other sectors...we can't do it' (interview, 8 September 2006). This view was empirically underpinned by recent research in the area that attests to the widespread failure of services to take account of population growth through migration (Polzer 2007). Thus, the municipality had been told that if they wanted more water, they would have to trade with other sectors, primarily agriculture, since they were the biggest users.

Dryland farmers

An estimated area of about 20 000 ha in the Nkomazi district is used for dryland farming, primarily maize and cotton (MBB 2000, cited in Waalewijn 2002). The Nkomazi Cotton Farmer's Association totalled around a thousand people (chairperson of Nkomazi Cotton Farmer's Association, interview, 29 November 2006), and there were also a number of dryland maize farmers. Whereas 'holding this thing' together referred to the individuals in the sugarcane schemes, there was a prevailing sense of disappointment amongst the dryland farmers who had tried to access water: 'They promised us that the dam would solve all our water problems. Now the dam's built, and it is full, but we are still not getting any water' (interview, cotton farmer, 7 September 2006). The dryland farmers that had tried to access water had done so through a variety of channels; some had gone to the Irrigation Board to ask for help, others had sent letters to DALA requesting water, and one told us he had sent a letter directly to the head office in Pretoria to ask for water. However, few of the farmers with whom we spoke knew about the representatives on the CMA that were supposed to speak on their behalf.

The reasons for scarcity were thus contested and were reflected in splits along sectoral lines, with irrigated agriculture, both commercial and emerging, pitted against the domestic users and dryland farmers, who were losing out. The emergence of a situation in which the commercial and emerging farmers shared common interests, and the discourses of the commercial farmers and extension officers was one of 'holding this thing together', while agriculture was pitted against other users was, to a large extent shaped by and mirrored in, inter-departmental relations. DALA was perceived by emerging sugar farmers to be quite flexible and lenient, with the chiefs approaching the head of the Technical Support Services to ask for water. The DWAF regional office, on the other hand, was taking an increasingly strict line towards both emerging and commercial farmers.

Inter-departmental struggle for allocative authority

The homelands were formally disbanded in 1994 after the demise of the apartheid regime, and the two-year-old agreement between South Africa and the KaNgwane homeland was rendered null and void. Although the agreement was no longer valid in strictly legal terms, it still provided practical guidance in matters of allocation between different stakeholders, reflecting the notion that 'law is whatever people identify and treat through their social practices as law' (Tamanaha 2000: 313, quoted in Oomen 2005: 31). Although the treaty (the rules of allocation) was officially defunct, it continued to be adhered to in practice, but became characterised by continuous negotiation and contestation.

In a letter to DALA dated December 2001, the then Minister of Water Affairs and Forestry, Ronnie Kasrils, writes that although the allocations made under the KaNgwane Treaty are 'technically not legal,' (as the treaty was agreed with what is now a defunct entity, the former homeland) they should nevertheless 'be allowed to continue'. DALA was still administering the bulk water allocation of KaNgwane. Although the existing uses were 'allowed to continue', DWAF still maintained the right to make reallocations as it saw fit. Thus, in order to meet the growing demand for domestic water, part of the low-assurance irrigation water allotted to KaNgwane was converted into high-assurance water for domestic use (see Table 4.4). The end result was that 14.4 million m³ were set aside for primary use, leaving 91.4 m³ of low-assurance for irrigation.[16]

TABLE 4.4 *Water allocations to the Nkomazi and Mswati, in million m³*

		High assurance	Low assurance
Mswati	Domestic	3.6	14.6
	Irrigation		
Nkomazi	Domestic	3.0	
	Additional domestic*		14.4
	Irrigation		91.4
TOTAL		6.6	120.4

* Converting low assurance to 11.4 million m³ high assurance

This did not go down well with the head of Technical Support Services at DALA. According to him, the DWAF knew very well that the 6.6 million m³ allocated in the treaty for high-assurance primary purposes was not going to be sufficient to meet the needs of the population of KaNgwane. The DWAF's proposal in the face of rising demands – to convert more of the low-assurance irrigation water earmarked for emerging farmers into high-assurance primary use to satisfy the domestic use – was questioned. In his view, this was not the way to go about it, asking '...why are they not taking this water away from the commercial farmers, or Eskom, rather than the small-scale farmers?' (interview, 23 June 2006).

There was another tussle related to the allocation of water, with respect to the Driekoppies relocation project. When the Driekoppies Dam was built, some of the people living in the dam catchment area were relocated. Several women's groups farming vegetables in the area, such as the Khutselani women's group, were provided with new water rights after relocation, but according to DALA, the Matsamo Tribal Authority was still 'owed' 300 hectares' worth of water-use rights that had been included.

DALA maintained a proactive role in administering water allocations, in particular to the sugar schemes. The impression given during interviews with the head of Support Services at DALA was that it was very much the department that was in charge in terms of water allocation issues. In his own words:

> When somebody wants to change the nature of water use, for instance, from agriculture to industry, they send it to us. Each application is a study in itself; you need to take account of the social impacts, and the economic impacts and also the long-term impact on the community in the area. It's really a full-time job just to do the applications.
> (Interview, 23 June 2006)

When asked who was responsible for handling applications, he responded that DWAF was the body that took care of what he termed the 'broad water management', in other words, distribution between different sectors, such as urban, rural, domestic, industrial and agricultural sectors. Whereas the DWAF regional office had made a block allocation available to DALA to administer in accordance with the treaty, the distinct impression given by the Technical Support Services during this interview was that DWAF's ambit was concerned more with 'general oversight' and 'broad management matters', rather than the nitty-gritty of the applications for water use and the nuts and bolts of site visits, which formed the basis on which an application was recommended for approval or not.

This perspective changed quite substantially from the first interview to the second. During the later meeting, he was quite clear that it was the regional office that had the ultimate responsibility, and that they (DALA) only dealt with applications that fell within the bulk allocation that they had from the regional office, this bulk allocation being the equivalent of what was allocated to the KaNgwane administration under the 1992 Treaty.

In the space of a few months, then, there was a palpable shift in the perspective that the DALA representative had on the role of his own department with regard to water rights authorisations. From being quite adamant that DWAF's role was restricted to one of being responsible for 'broad management', implying that DALA, mainly through the head of Technical Support Services, took care of the details associated with application handling, it was now clear that the positions had been more firmly demarcated, either by direct confrontation or by more subtle means. From the perspective of the DWAF regional office, DALA did not have any real authority over water issues. A senior engineer at the regional office put it as follows: 'They [DALA] are not really handling water use applications. They are only taking care of the enquiries coming from those lands that used to sort under the former KaNgwane government' (senior engineer, DWAF regional office, interview, 14 July 2006).

This impression of struggling for authority is reinforced by the reading of the minutes from the Mpumalanga Committee for Co-ordinating Agricultural Water (MCCAW) meetings. During the meeting held in August 2005, it was noted that: 'It was once again emphasised that DALA plays an important role in the whole process of evaluation on any application, with inputs from their side guiding the issuing of a licence within the framework of the National Water Act.' There is no role player/stakeholder with a 'veto vote'. In July 2006, the matter was brought up again in very blunt terms, and the first item on the agenda was to 'discuss the way forward, and reconstitute the meeting'. The note in the minutes simply reads, 'Applications for water rights must go directly to DWAF.'

Whereas the head of Technical Support Services at DALA had maintained relatively genial relations with the chiefs, and had sought to support their claims against DWAF for more water, the regional office on the other hand issued a blanket statement that 'there simply was no more water available', and that it had already been allocated 'equitably' enough. 'Close to half' of the water had been given to HDIs, according to the deputy director at the regional office (interview, 8 September 2006). However, he failed to mention the relevant demographics. In the area in question, about 120 commercial farmers occupied approximately 18 000 ha, while there were about 1 200 emerging farmers occupying roughly 9 500 ha. From the point of view of sharing amongst the groups, it may have appeared equitable, but if the perspective was quantity of resource per capita, the picture changed quite dramatically.

In the last week of September 2006, my research assistant and I went to the Upper Komati area to speak to the chairperson of the Upper Komati Irrigation Board. During that interview, he told us that

> there is still some 10 to 12 million m³ of water that has not been allocated. The government [the regional office] doesn't want to allocate it...they are doing soil surveys and ever more soil surveys, because they are reluctant to see the water come into the hands of the tribal authorities.
> (Interview, Upper Komati Irrigation Board chairman, 26 September 2006)

Some of the staff at the regional office were mistrustful of both the emerging farmers and the commercial farmers. Whilst the commercial farmers regarded themselves as making efficient and beneficial use of water, views at the regional office were far less favourable. As the deputy director put it, 'the guys are over-irrigating'; 'guys' referring to both the large and small-scale farmers.

> The soil is actually decreasing in quality, by over-irrigation. Very few farmers want to think about this. They just say 'give me water'...if we can cut 20 per cent of agriculture, it makes a huge difference. In a drought period, the crops won't drop the same way the water drops. If water drops by 40 per cent, the crops only drop by 10 per cent.
> (Deputy Director, interview, 8 September 2006)

There was a pervasive feeling of mistrust of what the farmers were doing, in particular the commercial ones. During an interview on 14 July 2006, with a senior engineer at the regional office who was in charge of the registration database, he declared that there was little reason to have any trust in the goings-on of the Irrigation Boards. During a later interview with him and a colleague, they stated that '...the big guys [are installing] more water storage...but we have little say in terms of the monitoring of the dams; that is the responsibility of the Head Office in Pretoria' (interview, 6 December 2006).

The stance of the regional office was clear: a desire to curb agricultural water use. Compulsory licensing was regarded as the primary means of achieving this aim. In the Deputy Director's words: 'Don't look to agriculture as your saviour. They should start building more factories...we are exporting too much raw material, and importing too much processed goods...Compulsory licensing will be a way to cut down agriculture' (interview, 8 September 2006). The view emanating from the DWAF regional office, that commercial farmers should be curbed via compulsory licensing and lowering assurance of supply, underscored the tensions between DALA and DWAF over the authority to issue water-use rights. The DWAF regional office discourse was effectively creating new subject positionings, basketing emerging farmers and commercial farmers into one category as 'wasteful' and 'inefficient'.

Trading water-use rights

Apart from the tussles between DALA and the regional office over administrative water allocations, the only other route for reallocating water was through trading. As noted in Chapter 2, the trading of riparian rights amongst users was not illegal under the previous Act, but the trading of quotas was. Nevertheless, in certain areas in the catchment, a brisk, largely illicit trade in permits had been taking place, particularly along the Crocodile River, where those in the upper reaches of the catchment had sold water rights to the tail-enders (several interviews with farmers and DALA staff). Some trade had also occurred on the Komati, again mostly from the upper to the lower reaches.

This informal trade was taking place largely among the commercial farmers. However, at DALA, the project leader of the Comprehensive Agricultural Support Programme (CASP) had tried to secure water rights for emerging farmers by approaching the Irrigation Board and asking them to identify any commercial farmers willing to sell their use rights.

> If there are projects [where] we need to purchase water rights on the open market, we'll do it...last year we bought rights for a 13 ha project; this year we are trying to get water rights for a 27 ha project close to the Komati River. They are demanding R25 000 per ha of water right. R25 000! They know they can ask whatever price they want, since it is the government who is the buyer. Then we have to apply to the department...there's so many checking and rechecking and cross-checking mechanisms in place. The system doesn't support development projects.
> (Interview, CASP project leader, 27 November 2006)

Thus, when the government acted on behalf of emerging farmers, it did so in a monopsonistic market, since it was the only 'customer' and therefore had to pay the asking price – which underscores the point made earlier that redistributing use rights through a market mechanism could be prohibitively expensive.

Another rumour that we frequently heard was that commercial farmers were selling off water rights on farms that were under claim. When we approached commercial farmers to talk about water reform, a frequent response was 'why do they want to do water reform? Can't they just wait until the land reform has been completed, and then all the water will have been reallocated along with the land?' What happened sometimes was that farmers with land under claim would sell off their water rights to reap the value of the water, at the same time drastically reducing the worth of the land.

Trumping intersectoral transfer

Although informal trading took place within the agricultural sector, there was little evidence of trade between sectors. DALA did not entertain any thoughts of engaging in intersectoral trade in water. The head of the Technical Support Services commented that

> they [DWAF and the developers] say we do not stimulate development as a department, because we do not want to give out water...they regard us as not stimulating for development. Developers want to buy water from agriculture and convert it into primary water use...and when we say 'sorry', they then go to the politicians and say that the development is going to generate employment and stimulate growth. But agriculture is not giving the water. (Interview, 23 June 2006)

He went on to explain that the farmers had invested in planning and infrastructure, they had made a great effort in order to utilise that water, so now why should they sell it? It was their right, they had worked hard for it, and they deserved to hold on to it. Thus, the potential efficiency gains to be made from a prospective trade – in allocating water to a higher value use – was trumped by the agricultural sector, drawing on Lockean principle of 'desert by labour' (Munzer 1990).

The little man against the State

Whilst DALA was fighting for control with the regional office, the commercial farmers were increasingly viewing themselves as victimised by DWAF. As the CMA representative for the commercial farmers put it, when we met him: 'Do you know what the biggest problem in this area is? DWAF. DWAF is our biggest problem' (interview, 11 July 2006). Since compulsory licensing had not yet been implemented, the only way anyone could obtain a licence was by converting one water use to another (for example, from agriculture to industry).

The Nkomazi/Onderberg Water Action Committee (NOWAC) was a temporarily-organised pressure group formed to defend the interests of the commercial farmers. In 1999, the group produced an extensive report, funded by the TSB, on the contribution of irrigated sugarcane to the Nkomazi area, as there was at the time a debate on the desirability of growing water-guzzling sugarcane in an arid region such as South Africa. The legal practitioner in Komatipoort, who acted as attorney for NOWAC, was quite an influential figure with many national contacts, and was also the former chairperson of the Water Tribunal. Her view of the situation was that

> the government blames politics all the time, but the main problem is that we are not able to get a licence out of DWAF...my clients go bankrupt while

they are waiting for a licence...It is taking them years – four years in one case – to process applications. That is unacceptable...now, what is reasonable for an application is about three months; that is what would be normal for a moderately complicated case. I don't really want to, but now I am advising my clients that they should start claiming damages for lost yields, etc., after three months...There are thousands of illegal developments going on, because you cannot get permission. Between the man on the ground and the policy-maker, there is a huge gap.
(Interview, 7 June 2006)

DWAF was reviled as a laggard. The attorney observed that many of the commercial farmers in their appeals noted that when farmers or foresters appealed against DWAF's decisions, the judges tended to be largely sympathetic towards the farmers and rule in their favour because, as she explained to us, to the judges the appeals represented the struggle of the 'little man against the State'. As the judiciary, it was part of their remit to function as a check on the State's power and supposedly, they were fulfilling this role in helping the appellants win their cases.[17]

Preparing for compulsory licensing

'They cannot force the river to look like they want to'
(Legal practitioner, Komatipoort, interview, 7 June 2006).

Validation and verification of water use

It was against this background of confusion and contestation that the preparations for the implementation of compulsory licensing began. Because the existing lawful uses had been entrenched in the policy, such uses would need to be *validated* and *verified* before the compulsory licensing process could begin. The eventual aim of the Water Allocation Reform was to convert all existing certificates of lawful use into licences, which necessitated a process to validate that the amount of water registered was actually used, and then to verify whether the registered use was lawful or not.

DWAF appointed a consortium[18] to do validation and verification in the Olifants and Inkomati Water Management Areas, with a timeframe from early 2003 to September 2005. The information that the team was to obtain was 'water information that the users may or may not have already supplied during the registration processes' (from a draft letter of information request to Irrigation Boards, 26 July 2005). There was considerable conceptual confusion around the terms 'validation' and 'verification'. Initially, the action of 'verifying' was taken by some DWAF staff to mean checking the 'truth' and 'correctness' of water use certificates, whilst 'validating' was taken as 'establishing legal force', that is, the validation of the implementation of article 35 processes (see Appendix C). The Director, Water Allocations, did his best to clarify the intended meaning of the terms, explaining that validation was concerned with whether the information provided for registration was correct; whereas verification was to be carried out in order to confirm the legal status or lawfulness of water use. In the *Guide to Determining Existing Lawful Use* (DWAF 2006), 'validation' and 'verification' are explained in the following terms:

> Validation confirms how much water the user was actually using in the qualifying period, how much they said they were using (or registered), as well as how much they are currently using. Verification determines the extent of existing lawful water use. In effect, this determines if any previous laws would have limited the use in the qualifying period.[19] If not, the use in the qualifying period is lawful...Directives may be issued to water users to stop that portion of their current use that is unlawful. A guideline document which outlines how this should be done is also available to guide regional offices in taking appropriate steps to address unlawful use. Verification of existing lawful water use is required before implementing the compulsory licensing process.

The regional office in the Inkomati sent out a letter (dated 26 July 2005) to the Irrigation Boards prior to the commencement of the validation and verification project, in which the terms were explained as follows:

> By validation, we mean the checking that the information registered on our database is actually being irrigated in the field, or is scheduled...
> By verification we mean the checking of the lawfulness of the water use taking place, i.e. is it scheduled or is there a permit or other authorisation for the use?

It further explained the intention of the validation and verification process as 'to collate the following information for each property where water use is identified':[20]

- the cadastral data (which includes extent, value and ownership) relating to the property;[21]
- the registered use (volumes, irrigation systems, crops and hectares);
- the use in the qualifying period (volumes, crops, irrigation systems and hectares);
- the current use (volumes, irrigation systems, crops and hectares); and
- whether any previous legislation would have limited the historical use.

Thus, the efforts to validate and verify water use were linked to cadastral data, and thereby to property ownership, which rendered the process fraught with difficulties in terms of validating and verifying use in 'tribal' areas. There was heavy emphasis on the need to get accurate cadastral data, with the report arguing that, 'the importance of clean, accurate cadastral data is of extreme importance to the verification process. Impacts on the project are significant.' Therefore, although land and water rights had been separated, in practice it was necessary to re-link them in order to determine whether existing users were 'lawful'.

Uncertainty and dynamics

There were two main challenges with the validation and verification project. One was the problem of getting reliable data on water abstraction rates. The other problem was that land was a dynamic asset, which was continuously changing hands. Since water use in the previous Act was dependent on land ownership, it would be difficult to pin down the lawfulness of existing uses given the dynamics of land transactions.

The difficulty of determining water abstractions

With respect to determining water abstractions, estimating the water use that had taken place in the qualifying period was a difficult task. The team relied on a combination of GIS, satellite imagery (primarily LANDSAT data), and the SAPWAT model to estimate crop-water requirements. Since historical water use was rarely measured, determining use in the qualifying period represented the most challenging element of the project. Remote sensing and LANDSAT data were used to identify areas of irrigated crops and plantation forestry. The cadastral data were then applied as a 'cookie cutter' to determine the area under irrigation on each property.

The main challenge with the satellite approach was that it was difficult to discern crop type. The timing of over-flying also mattered, as fields might have been lying fallow during that specific period. Once the areas under irrigation had been determined using remote sensing data, the SAPWAT model was applied to estimate the quantities of water used for irrigation. However, the difficulty in discerning crops carried over to the SAPWAT approach as well. Since it was primarily a means of estimating crop-water requirement, the type of crop was a crucial parameter, so a mistakenly identified crop would yield irrelevant results. Moreover, results generated by SAPWAT vary greatly depending on the attributes it is fed – for instance, crop characteristics, crop requirements, crop type, planting dates, soil type, infrastructure, irrigation practice and slope gradient – and the variable interpretation of operators. Even estimates of the model's error margins differ substantially. One internal DWAF source reports SAPWAT as having an error margin of up to 40 per cent, whereas others quote figures as high as 60 per cent. The estimates of current water use generally followed the same procedure as that described above, but with additional trips into the field in order to compensate for the errors generated by SAPWAT.

The DWAF regional office in Nelspruit perceived SAPWAT as problematic, not least due to scarcity of expertise:

> SAPWAT is a problem, we do not have enough people with the necessary skills to interpret the data…you can come up with a lot of answers…you have to interpret the answers, but you need to be an irrigation specialist. It's a problem.
> (Interview, senior engineer, DWAF regional office, 14 July 2006)

Because SAPWAT merely estimates the crop-water requirement, it only offers an estimate of the optimum water application rates for that particular irrigated crop, rather than the actual use. The model tends to overestimate water abstraction rates, since the team found that irrigators routinely apply less than the crop requirement in order to save on electricity costs for pumping. The practical upshot of this was that using the SAPWAT model could overestimate the water use in unscheduled areas. This also had implications for the remote sensing data, as the project team argued that aiming for high accuracy data was rather pointless, considering that the inaccuracies of SAPWAT led to distortion anyway.

This resulted in considerable confusion with regard to how much water was perceived to be actually available, because the DWAF regional office insisted that there was a situation of over-allocation; and this reason was quoted when refusing to issue water-use rights to emerging users. The estimates of the project team on the verification and validation of water use determined that there was, in fact, more water in the river than the regional office people had calculated, something which took the latter by surprise. As one of the people present at a project meeting exclaimed, 'How the hell did this happen?' Speculation ran that either there was more water in the river than previously thought, or that people were using less water for irrigation, for example, responding to periods of low flow in the river and thereby irrigating less; or that SAPWAT had overestimated water use.

A particularly tricky issue was determining the past and present use of the 'non-scheduled users', that is, those who had been utilising water in terms of their riparian rights, but were not included in a rateable or scheduled area. Because they were exercised as correlative rights in regions outside Government Water Control Areas (GWCAs), users were primarily exercising their riparian rights to a reasonable share of normal flow (see Chapter 2), and information about these users would have to be gathered through interviews and field trips.

The consultants had needed people to go out into the field and do the interviews, and had contacted the Lowveld Agricultural College to ask if there were any competent students who could undertake this work. According to my assistant, who had been part of the team, they recruited four students, two female and two male. Whereas he and the other male student had been provided with a car to get out to the farms, the team had decided that it would be too costly to supply the female students with a vehicle as well, and they were therefore left mostly in the office and asked to do telephone interviews. In my assistant's words, 'they [the farmers] could tell them anything; there was no way they could actually know that what they told them [the female students] was true, since they could not see the farm' (research assistant, personal communication).

In conversation with a senior lecturer at the Lowveld College of Agriculture in Nelspruit, he related an incident in which a group of scientists had come to demonstrate models and formulae for measuring water abstraction rates and flows. 'When the scientists came into the room, the farmers just rose up and walked away...Farmers and scientists live worlds apart. They [the farmers] live by rule of thumb' (interview, 23 May 2006). He went on to remark that although the new policy made sense, there was little chance of DWAF getting the information they sought, pointing out that

> DWAF didn't manage in 50 years to get an overview of existing irrigation systems, so how do they think they are going to be able to get information on all kinds of irrigation systems now? Not in a thousand years.

From the farmers' point of view, the project was too engrossed in its own ideas of scientific detail to be of any relevance.

> They wanted to know everything in detail, for example, how many mangoes you grow. Thing is, I'm not necessarily going to stick to that specific regime what they asked about at that specific point in time – things are going to change. The lady [the project leader] was sweet enough, but they wanted very specific data, very accurate data.
> (Farmer, Lomati River, interview, 24 May 2006)

The centralised, scientific approach to estimating water use also influenced the perception of institutionalised collective action, as explained below.

During the registration process, many Irrigation Boards registered as single users in the WARMS. This concurred with the department's official view that such associations would be expected to engage in some degree of self-policing. This view was changing, however, as the issue of the single registration of Irrigation Boards with WARMS caused the consortium something of a headache and there was a considerable degree of conflict and confusion between DWAF headquarters and the project partners. The DFID consultant wanted all irrigators identified and registered, and claimed that the team had used incorrect data and had not acted on the instructions it was given regarding the inclusion of the spatial data. The consortium was asked to find out which Irrigation Boards and Water Users' Associations had registered with WARMS, whether they had registered all their members, and then define the number of double registrations, that is, where an individual was registered with the Irrigation Board but also registered separately as an individual with WARMS.

Thus, there were profound problems with the project, in terms of diverging views and interpretations, but also in terms of the actual way in which the assessment was carried out in the light of the many uncertainties involved.

Messy legal context

Linked to the difficulties of obtaining 'clean' up-to-date cadastral data, was the growing realisation that the dynamics of continuous land transaction influenced the legal status of water rights. Where land had been subdivided, it was often not possible to determine to which plot the riparian water right was now attached; similarly with consolidations. The dilemmas inherent in using a relatively static database to mirror dynamic land relationships have been extensively documented in research on land titling efforts across Africa (see, for example, Benjaminsen & Lund 2003), but the implications for water rights have seldom been explored. In addition to land dynamics, informal trade in water rights contributed to complicating the picture. According to the previous legislation – the 1956 Act – users were permitted to trade riparian rights (the fair share of the normal flow) as they saw fit, but not rights attenuated through government-issued permits or quotas (see Chapter 2: 21). Illicit trade had nonetheless taken place with the covert consent of local Irrigation Boards, to which examples from the Crocodile catchment in the Inkomati attest (Bate & Tren 2002). Another aspect of the need to attach water rights to land was the uncertain nature of land tenure in tribal areas (see Chapter 2).

Outcomes

For the Inkomati and Olifants catchments, the final report on the validation and verification project found that in 67 per cent of cases it was impossible to determine the lawfulness of water use. Problems arose from properties having either been split up or consolidated, without water rights having being adjusted to reflect this, which would happen through, for example, informal trade having taken place. In only 16 per cent of the cases were the assessors able to conclude that water use was lawful, with the remaining 17 per cent being characterised as possibly unlawful. The report argued that the time and resources devoted to this project – it had taken more than

22 months to complete – were excessive. This, in short, was a rather disheartening result for the Directorate, Water Allocation.

The attempt to shift rights relations from correlative user–user (as in the riparian system) to State–user (as in the administrative rights system) by means of registration and technology largely foundered. Drawing on Jasanoff's conception of co-production, which essentially entails 'thinking about natural and social orders as being produced together' (Jasanoff 2004: 2), helps shed light on the boundaries between the 'public' and the 'private', and how science and technology mediates these boundaries. The unscheduled users are partially outside of the technological scope of government control, and therefore difficult to define, verify or curtail. Through the discourse on existing lawful uses, such uses have been entrenched as property rights (see Chapter 2), and the inability of the scientific and technological apparatuses to demarcate their content renders such recognised rights of use stronger than licences. They enjoy the status of a right, but with little scope for attenuation.

Since SAPWAT was so inaccurate, DWAF realised that 'it will be very difficult to prove a water use that exceeds the allocated verified volume in a court (or tribunal) with the presently available mechanisms and legal framework in the NWA' (DWAF internal source). Garduño and Hinsch (2005) point out in their report that the support of compulsory licensing requires far more sophisticated modelling techniques, and it would take 10 consulting teams approximately 20 years to complete. Interestingly, the proposed way out of this dilemma was to argue that unlawful use was inefficient:

> If users may have registered more than the SAPWAT requirement, [they] may argue that they used more [than] this during the qualification period. As it may be difficult to prove that this was not the case (and hence to prove this use as unlawful), the use could be flagged as inefficient and be considered as such during the compulsory licensing process.
> (DWAF internal note)

Thus, the early discourses of existing lawful use as 'efficient and beneficial' were undergoing scrutiny. The failure of the State to curb unlawful use was, in a sense, closing the circle. Since they were not technically or scientifically able to pin down unlawful use, the 'weapon' resorted to was again a discursive one, in labelling users as 'inefficient.'

A couple of months after the report on validation and verification had been finalised, a workshop on the impacts of trade liberalisation on sugar growing in the Komati was held in Pretoria, following the publication of a report (Lorentzen & Cartwright 2006) on the subject commissioned by the World Wildlife Fund (WWF). The workshop was targeted mainly at policy-makers and interested academics. At that workshop, the DFID consultant went public with the results from the validation and verification of water in the Inkomati. In essence, he stated that water use in the Inkomati is still authorised as existing lawful use. However, any expansion of water use since October 1998 is unlawful. Irrigation in the Inkomati has expanded since this date. This is probably unlawful since it was not allocated. DWAF has been holding back 18 000 ha of water from emerging users because of 'insufficient water', but we know there has been 17 000 ha of unlawful expansion.

So, the news was out. At lunch that day, the Director: Water Allocations remarked that people who thought the compulsory licensing process was going to go ahead smoothly 'were in for a shock'.

These events attested to the fact that DWAF headquarters was realising increasingly the shortcomings of the scientific approach to determining lawful use, and was beginning to outline alternative means of coping with this, through, for instance, labelling existing users as 'inefficient' (in direct contrast to the earlier focus on ELUs as efficient and productive). The people within the Directorate: Water Allocations were growing impatient with regard to providing water to the HDIs; they wanted to speed up the reform process and needed to find new ways of doing so in the light of the increasing difficulties.

One solution was to further develop the concept of General Authorisations (GAs). The idea of GAs was not new – it was one of the four categories of use rights defined in the Act (see Chapter 2: 37). However, at that time, it was viewed primarily as a means of lessening the administrative burden associated with licensing, and intended for use in unstressed catchments that did not yet need to go through the compulsory licensing process. But as the water allocation reform dragged on into its seventh year without much progress, the idea of developing GAs as a tool to be integrated into the compulsory licensing process slowly formed, although opinions on its potential differed within DWAF. Nevertheless, a team of consultants from Cape Town-based Ninham Shand was asked to develop the idea further, and in November 2005 an inception workshop was held for interested stakeholders. They comprised the regional directors as well as representatives from other interests, such as IWMI and officers from the Department of Agriculture and Land Affairs and various financial institutions. GAs were seen as sitting in the 'grey area' between Schedule I users – water for small-scale domestic use and non-commercial purposes – and licences, which were viewed as providing access to water for commercial purposes. The view that emerged at the inception workshop was that GAs should be reserved for small-scale use by people with their own abstraction technologies, and that water should be allocated primarily to users that were able to take up the water and use it productively, not simply because the users are HDIs. Thus, the same logic obtained here as had applied in the case of licences – equity hinged on efficiency or productivity. There was no mention of infrastructure in terms of enabling access to water for HDIs.

The CMA, meanwhile, felt that it was being held on a tight leash. Though some people within DWAF had envisaged an expansion of power to include the issuing of licences (Schreiner & Van Koppen 2002), there was considerable tension with respect to the bounding of the CMA's powers vis-à-vis DWAF. Some of the board members were of a mind that the CMA should be given the authority to implement the compulsory licensing process and issue licences (interviews, several CMA reps, July–September 2006). However, there were clearly tensions with respect to the nature of power-sharing and lines of authority between the CMA, the regional office, and DWAF, with one board member in particular arguing that DWAF was consciously holding them back and denying them the financial resources to operate fully (interview, 11 August 2006). This resonates with the assertion in the introductory chapter that the decentralisation process carried with it an inherent tension with respect to allocating authority.

Impasse

After the validation and verification project had been completed, the GA process picked up again. In October 2006, another workshop on GAs was held in the Inkomati. Present at this workshop were staff from the DWAF regional office, as well as representatives from DALA and a few external consultants. One of the engineers from the regional office observed that although GAs might potentially lessen the burden of administration, they created a new problem: one of prioritisation. If they received a number of applications for GAs, and if these could not be met with the available water, what guidelines should they adhere to? A proper socio-economic analysis would be needed to make decisions on the best opportunity costs of water, which could be time-consuming.

This point echoed the views of the senior engineer and designated 'Water Allocation Reform Champion'[22] at the regional office, whom I interviewed a couple of weeks before the workshop. He argued exactly the same point, saying that '...the other scary thought about GAs is that this one guy can get as much as he likes, but not that other...Who do you give it [the water] to?' (interview, 8 September 2006). He continued (my emphasis):

> It's a nice idea...but I'm very scared of the term 'General Authorisations'. It implies that you don't need *control*. If 5 per cent of the water [is] set aside for General Authorisations and you don't monitor it, you don't authorise it, you don't keep records of it...it means you don't have *control*.

He said that there was a study underway to assess the water availability in the system, which was planned to take about two years, and added:
> if the politicians say they can't wait to reallocate, then there's going to be a huge risk. I have a major problem with that. To do [compulsory licensing], without having all the data is very risky. You cannot reallocate without the necessary information. They haven't even set the international requirements yet.

The Director: Water Allocations had a different view, however. He seemed rather exasperated with the notion that the regional office wanted more hydrological studies to be undertaken:
> Hydrologists, modellers, they want great detail; you can't, you've got to take some risk…It takes them too long…How much should you study a system before making a decision?
> (Director: Water Allocations, interview, 31 October 2006)

There was increasing uneasiness within DWAF as well. The DFID WFSP project was nearing its end and funds for further reform progress were therefore drying up. Key officers who had been part of the process were leaving in droves, seeking greater job satisfaction and higher salaries in the consultancy sector. This drain on already thinly-stretched resources emphasises the importance of 'second order' capacity in terms of carrying through reform. Thus, the water allocation process was largely left in an impasse.

This chapter has shown how the subject positionings at the local level were far more fragmented and dynamic than the policy-level discourses. Emerging and commercial sugar farmers were pitted against other interested parties, including domestic and urban users. Emerging farmers, although made out to be a success story at the national level, were struggling, and the regional office was refusing to issue any more water to emerging farmer projects. Instead, the regional office's Deputy Director portrayed most farmers as 'water wasters' and sought ways to curb agriculture. This way the shift from riparianism to administrative authorisation opened up a space for the ensuing struggle for the allocation of authority between DALA and the DWAF regional office. Whilst other potential users were increasingly disappointed, the commercial farmers associated with a discourse of 'the little man against the State', drawing upon the moral high ground of victimhood. Further, the process of preparing for compulsory licensing in the area was put in a quandary by the difficulties of accurately defining water abstractions and the dynamics and 'messy' nature of land holdings, which left the reform efforts in a temporary impasse.

Notes

1. The other is the Olifants.
2. Government Notice 124, 1972.
3. More than 20 000 Mozambicans have returned across the border following the spate of violent xenophobic attacks that erupted in Alexandra in May 2008, and which were directed at Zimbabweans, Mozambicans, Malawians and other African foreigners in South Africa (*Mail & Guardian* 28 May 2008).
4. One cubic metre of low assurance supply equals 0.794 m^3 of high assurance supply.
5. As a result of the IncoMaputo agreement reached in 2002.
6. The department was renamed the Department of Water Affairs in May 2009.
7. As of February 2009, there are 3 established CMAs.
8. http://www.award.org.za/home/
9. http://www.geasphere.co.za/indexa.htm
10. LRAD was initiated by Thoko Didiza in 2001, shelving the former Settlement/Land Acquisition Grant (SLAG) programme. The LRAD scheme provides an initial R20 000 grant on a sliding scale and has been criticised for moving away from the pro-poor focus of the SLAG programme to align with GEAR policy of emphasising entrepreneurship, favouring those with capital.
11. Afrikaans for 'pick-up truck'.
12. Ratooning is the practice to regrow the sugar plant from its own base.
13. PTOs are issued either by a communal or tribal authority on state land within the former homelands. A PTO provides the bearer with a recognised right of occupation and utilisation of an identified portion of land, but the land remains the property of the state. PTOs are not recognised for purposes of security by financial lending institutions. This situation is the cause of low levels of investment by the private sector in such areas. In 1994, however, with the inauguration of a new government and later a new constitution, land was made a national competency. In terms of the administration of state land, this function was vested with the Minister for Agriculture and Land Affairs. Currently, most of the former homeland areas (South African Development Trust – SADT land) is vested with the national department. In strict legal terms, since 1994, only the minister has had the authority to issue a valid and legally binding PTO. Furthermore, a PTO provides the holder with a recognised, albeit informal, right to land. These occupational rights are protected, however, in terms of the Interim Protection of Informal Land Rights Act (No. 31 of 1996) (IPILRA).
14. Following the Land Acts of 1913 and 1936, there were special land delivery procedures enacted, known as Proclamation R293 of 1962 for proclaimed urban areas, and R188 of 1969 for rural areas (Fourie 2001). R188 settlements were administered by tribal authorities, who could allocate lands for residential, commercial or agricultural use by tribal members. Residents of these settlements occupied land through the issuing of a Permission to Occupy (PTO), which granted the holder utilisation rights over the land. Generally, these settlements were unsurveyed and unplanned and have continued to grow on an incremental basis as the need has arisen for more land for residential use.

15 In an interview on 27 November 2006, the chairperson of the former Elands River Irrigation Board gave us his opinion of why transformation was so slow: '...we were faced with some difficulties, how to get these different sectors, some very big, some very small, without allowing the bigger ones a major say. I decided [to] only allow associations or institutions like municipalities to become members...it makes the membership number much smaller and much easier to control. While this process was going on, the law regarding WUAs was changed [so that] we had to also include membership of the rural population...and also farm workers, which we did. Each municipality had a person employed who is in charge to look after the needs of the rural population, but it isn't easy. I try to do my best in adhering to what the policy is. But they [DWAF] have messed everything up, by continuously shifting ideas, policies. The other Irrigation Boards [in this area] didn't find it necessary to follow up under these messy conditions...the department just sits there and doesn't listen to you.'

16 When assurances are converted, this will also impact on volumes. If a given volume of low-assurance water is converted into high assurance, that volume is necessarily reduced.

17 These particular cases were not officially available through the Tribunal, and my attempts at gaining access to them via DWAF staff and the legal practitioner were, unfortunately, unsuccessful.

18 The consortium consisted of Senqu Consultants, GeoTerraImage, Siphumelele, MHP Geospace, CPH Water, CSIR and Vikna Consulting Civil and Developmental Engineers.

19 The qualifying period was defined as the two years prior to the promulgation of the National Water Act 1998.

20 This does not include properties using water under Schedule I of the National Water Act.

21 A 'cadastre' is a public record, survey, or map of the value, extent, and ownership of land as a basis of taxation.

22 DWAF headquarters had identified a 'WAR champion' at each regional office, who would assist with implementing the reform. However, the black WAR champion initially instituted at the Inkomati regional office had been replaced by a white one when the former was posted to another region, for reasons unclear to me.

5

Conclusions

THIS FINAL CHAPTER SYNTHESISES and ties together the main points made in the book so far, and offers some reflections on the wider implications of the South African reform efforts. What has the study yielded with respect to understanding policy reform processes and allocation of use rights to water resources and how these mesh with on-the-ground perspectives? And why does it matter? The following sections tease out some of the main insights. Shifting political conjunctures and the language and terms used in policy have an impact on how use rights are conceived. Particular ways of portraying water use and users are employed rather than others, giving rise to 'privileged accounts' that emphasise certain features and suppress others, highlighting how ideas about scarcity, equity and efficiency came to be presented, with the existing users seen as being beneficial and the process of redistribution as risky. Moreover, although land is a hugely important part of the water reform process, little effort was made early on to link these two reform processes, or the related issue of the authority of the chiefs, in any coherent way. A core theme is the fact that water is hard to control, due to the combination of inherent uncertainty and variability in hydrological systems and a paucity of technological means and capacity to measure flows and abstraction rates with any great degree of accuracy. The final sections comment on the recasting of the Water Allocation Strategy in late 2008, and how attention has since come to centre on the Strategy for Growth and Development launched in 2009. Reflections are also offered on the political shifts and future prospects of South Africa's water policy, as well as wider implications.

Drawing it all together

South Africa's turbulent history, and recent efforts to redistribute natural resources such as land and water, have been difficult and drawn out, as anyone familiar with the land reform process will readily acknowledge. Debates among different interest groups, and competing discourses of what development path South Africa should follow, have been dominating the political landscape ever since independence. Change and reform do not come easily. As Chapter 2 described, although the South African Constitution was fairly radical in many respects, among other things recognising water as a human right, it still retained the so-called 'property clause', which held that existing property rights should be protected. The ANC's initial Reconstruction and Development Policy, which was geared towards increasing social welfare and the redistribution of resources, rather quickly gave way to the much more neoliberal-oriented GEAR policy. Concomitantly, the Black Economic Empowerment policy surged ahead. This was intended to offer black individuals an entry point into business, and was essentially meant to be an affirmative means of trying to compensate for past discrimination. However, many have argued that these efforts have merely replaced the old order of a race-based society with a new class-based one, in which the wealthy elites don't care much about the masses still stuck in poverty. Those who feel that the ANC is taking too long to fulfil its promises to the poor are getting increasingly restless in the face of the mounting gap between the political and business elites and poor people. This emphasis on protecting existing property rights, and the neoliberal turn in terms of prioritising business interests, was quite clearly reflected in later drafts of water allocation policy, where the emphasis came to rest on protecting the 'existing lawful uses' (ELUs) and favouring those users who apparently made the most productive use of water. These particular political constellations, therefore, exerted a considerable influence on the policy discourses.

That politics influence policy is neither a surprising nor a novel observation. More interesting, however, is the observation that the way in which particular users are portrayed in policy, and the assumptions on which these portrayals are based, also contribute to shaping water rights and allocation. When water is vested with the State, the State retains the authority to determine who should receive water-use rights, and on what basis. A central argument of this book is that policy narratives – that portray existing and potential users of water in certain ways – profoundly determine such rights. This contrasts with conventional legal doctrines, such as riparianism, where users derive water rights based on their land holdings.[1] The nature of administrative use rights distinguishes water rights reform from land reform in two important respects: i) there is a considerable degree of uncertainty in terms of measuring and predicting water flows; and ii) water use is relative – one person's use will affect other potential users.

At the heart of water is *uncertainty* and *unpredictability*. Whereas land is stoically stable, water is a fluid and fugitive resource – a plot of earth won't vaporise beneath your feet; it won't trickle between your fingers and percolate the porous soils; it won't rush with sudden haste to join the seas. As has been demonstrated in this book, water resources are hard to measure and monitor with any great degree of accuracy, particularly given the capacity constraints prevailing in South Africa. The stretching of state capacity, in combination with an overly complex implementation strategy contributed to slowing down the whole reform process. The inability to determine rights and to monitor and control illegal abstractions is not only due to the characteristics of the resource itself, but also partly a result of capacity constraints, which is not made easier by the fact that it has proven hard to fill vacancies within the Department of Water Affairs and Forestry. The sharing of roles and responsibilities, particularly between DWAF headquarters, the regional offices and catchment management agencies (see Chapter 4: 90–91 and 129) is characterised by a lack of clarity, which has not aided the efforts to implement an overly sophisticated set of policies.

Water use is also *relative*. If a bucket of water is taken out of a flowing river, this means that there is a little less left, an opportunity reduced, for someone else further downstream. Riparianism is often misconstrued as a doctrine of 'private' water rights, but this is not the case. It is, in fact, a 'closed commons' (see Chapter 2, p 19), and the exercise of such rights depended on the idea of 'reasonable use' (see Chapter 2: 19) and was *correlative* to that of other users. Riparianism thus implies that riparian landowners engage in the management of water as a common property, each individual's use being in principle relative to the use of others (Rose 1994; Tisdell 2003; Backeberg 2005). The doctrine of riparian rights was based on the acquisition and ownership of land adjacent to a river, which resulted in a highly skewed distribution towards the white minority due to the discriminatory Land Acts. Prior appropriation, on the other hand, grants exclusive use rights to the most senior user, and the doctrine has traditionally been considered to be more suitable for areas where flows are variable. Prior appropriation, then, represents a system by which the individual user's rights are *not* correlative to the extent that they are in the riparian system, and therefore prior appropriation systems have been considered to represent 'private rights' to a far greater degree than riparianism. However, this doctrine was never introduced in South Africa (see Chapter 2: 20). Instead, with the introduction of the 1956 Act, some riparian rights were attenuated through the establishment of Government Water Control Areas (GWCAs), where the government imposed caps on riparian use through the determination of quotas, defined either as flow rates, or as volume per hectare. These quotas were adhered to largely at the discretion of the Irrigation Boards, since it was difficult to monitor actual use. The result was a mixed system of water-use rights arising primarily out of land acquisition, but where some of these rights were attenuated in areas of particularly high demand. Permits could also be issued to non-riparian users, thus facilitating the inclusion of urban and industrial interests. The 1956 Act represented a gradual shift, then, to more State authority, which culminated in the 1998 Act.

What the shift in legislation in 1998 achieved was to change the nature of relational water use from the riparian correlative user–user relationship with *ad hoc* intervention from the State, to a predominantly State–user relation. In this system, licensed water users are first and foremost accountable to the State, rather than to other users. This shift from riparian correlative use to a State–user relationship created an unprecedented space for the emergence of narratives on distribution, or *allocation discourses*, at the national level. How should the State deploy its authority in terms of allocating water to users in order to redress the existing skewed distribution? Chapter 3 showed how the reform initiative came to focus not so much on the redistribution of resources as on environmental concerns and the need to close the backlog in services. This gave rise to a pervasive vagueness and lack of clarity on resource redistribution. Many of the key terms and concepts – such as licences – were still quite vague and perceived in different ways by different actors. Chapter 3 showed how different narratives fed into this ambiguity, conceptualised as the *livelihoods* discourse and the *institutionalist* discourse.

The institutionalist discourse emphasised the idea of use rights as providing security to facilitate economic growth, but mainly in terms of the existing users, and cautioning against a confrontational approach to redistribution. The institutionalist discourse tended towards interpreting licences as providing more authority to users – in that users were now able to exert allocative authority through trading – but was wary of the uncertainty created by the State's ability to impose conditions of use.

The livelihoods discourse, for its part, though amenable to the idea of licences as facilitating reallocation, viewed the State's ability to impose conditions on users through licences (and the potential for the exclusion of users less familiar with the language of bureaucracy) with scepticism, and likened it to a form of continuation of colonial oppression. It emphasised the authority exercised by the State in terms of defining and allocating licences, and expressed the fear that it would exercise this authority in a neo-colonial fashion. The livelihoods discourse was keen to revive the notion of customary rights, which were perceived as inherently more equitable.

A key issue in the reform efforts was the further deepening and strengthening of the Water Act's concept of existing lawful uses by means of privileged accounts (see Chapter 3: 64), making associated users out to be productive and beneficial, and stating that it was in the public interest to retain such rights. Existing users had been defined through the 1956 legislation, which made a distinction between private and public water, where groundwater was considered to be private. In contrast, river water was public and governed through riparianism. Although the principle of riparianism was codified in law, the term *reasonable use* gave wide latitude for contextual interpretation, and in some cases, local norms emerged determining what was reasonable, such as the Lomati quota of 8 500 m³ per hectare. This volume was not stipulated through a government quota, but evolved as the local norm of usage (see Chapter 4: 107). Riparian use rights, then, are 'mushy' (Rose 1994b: 166) in the sense that the contents of the use right are determined by contextual factors – the reasonable use dictum, based upon the principle of proportionality, is dependent upon the seasonal nature of river flows, and of other users' needs and demands at certain times. These aspects serve to fashion the nature of the use right at any particular moment in time, and are monitored by forums of water users, with local water courts acting as arbiters. Thus, there were certain checks and balances in place to ensure that the principle of reasonable use was adhered to. Although these checks and balances were by no means watertight – excuse the pun – they at least served as constraints. Breaching the trust by stealing water would not only result in punitive action being taken, but also in public disgrace. The major problem with these institutions, however, was that they were almost exclusively white.

With the coining of the term 'existing lawful uses', a new set of use rights was effectively created. Through the creation of this category, user–user correlative rights had to be transformed into State–user relative rights, with the State – through the Department of Water Affairs and Forestry – now taking it upon itself to determine the content of such uses with a validation and verification project (see Chapter 4: 121 and the section below). Existing lawful use, then, became a category that brought the various existing users – both the *pure* riparian, and those attenuated through quotas within the GWCAs, under a new umbrella of *fortified* rights. Fortified, because existing lawful uses, by fiat, could not be conditioned in the same way as licensed use, and because the arbiter role of local water courts had been removed, thus effectively reducing the former checks on excessive use. Existing lawful use was thus exempted from the State-imposed conditions of licensing. This *ex post* creation of lawful use rights reflects what Bromley eloquently argues (2004: 26), namely, that the conception of property rights is 'not something that is known to us axiomatically – something whose essence is clear to us through intuition or introspection'. Rather, the idea of property rights is created, or 'arrived at'. Though Bromley takes various legal court cases as his point of reference, it is equally insightful to see how the concept of property rights is arrived at through the formulation of a policy of allocation.

The categories created at the national level did not resonate at the regional/local level. Rather than the neat dichotomy of ELUs and HDIs, the situation was a more fragmented and dynamic one, with the emergence of an alliance between the established commercial farmers and the emerging farmers characterised by patronage and paternalism. The emerging farmers were constrained by lack of storage, lack of flexibility, and little control over water, and added to this were the informal processes of land access, through which emerging farmers had come to occupy plots in the schemes that had little to do with individual ability and more to do with chiefly relations. The white commercial farmers retained control over their water rights by means of their own storage dams and exercised their clout in terms of actively participating in Irrigation Board decisions, where the emerging farmers had little real power. Although the commercial farmers expressed an interest in seeing the success of black sugar farmers, this did not imply that they would willingly share their own water allocations, basically framing the difficulties experienced by emerging farmers as a management problem that could be outsourced. Other users, meanwhile, saw the sugar farmers – both the commercial and emerging – as water wasters, a view that was mirrored in the ongoing contestations for authority between the DWAF regional offices and DALA. The commercial farmers regarded DWAF as a culprit and themselves as 'victims' of the attempts to monitor and control farmers' water use. At the time, the State effectively tripped itself up with the retention of ELUs, which involved an unrealisable reliance on scientific certitude to determine the extent of lawful and unlawful use. DWAF's continued attempts to find alternative ways around this dilemma and achieve increased equity saw the emergence of the idea of using General Authorisations more proactively to 'ring-fence' water for redistribution, but this was fiercely resisted by the regional office at the time, and led to difficulties in bringing the process forward.

Emerging insights

What kind of realisations and insights have emerged, and what implications do these have? A key insight was the idea that the influence of the IWRM paradigm and its emphasis, particularly the ideas of holistic and decentralised management, contained an inherent paradox. On the one hand, this paradigm emphasised increased participation and a decentralised institutional structure where decisions should be taken at the lowest appropriate level. On the other, it argued the case for maintaining central oversight of the resource as a whole. The transition from a doctrine of riparianism implied a shift in allocative authority, which transformed water-use rights from a system of user–user correlative rights to State–user relations, a shift that necessitated the identification and registration of all water users through the establishment of the Water Authorisation and Resource Management (WARMS) database. The shift thus required and depended on a technological and bureaucratic basis, in other words, on a database that would contain information on all the rights holders to enable the State to exercise oversight, and the necessary bureaucratic and scientific infrastructures to manage, monitor and enforce this system of rights, through the use of modelling technology, GIS and remote sensing. The State's transition to a regulatory role of overseeing rather than engaging in active management does not detract from the fact that it was the last port of call and retained the final say in the issuing of licences. It had the ultimate allocative authority, in contrast to the situation under the previous legislation, where such authority depended on the possession of riparian land. This shift in control, then, created space for policy actors to define the principles of allocative authority, and the conceptions of rights and users were actively formed through the narratives that emerged around these allocation principles. Examining the assumptions of these narratives became a way of understanding how use rights were conceptualised and what arguments underpinned particular conceptions. Story-telling is not only a feature of the individual voicing claims on property, but also a feature of larger policy processes. The discursive representations of particular categories of users – such as existing lawful users – effectively cemented their existing use rights, but removed the checks and balances inherent in the correlative system of riparianism.

How discourses shape water rights

The book's main argument is that understanding how water rights are formulated and made to work is enhanced through exploring the ways that use rights are construed through policy language. Water policy reform is not simply a tabulated, neutralised exercise of applying certain given principles, but a protracted struggle for meaning. Policy discourse itself actively shapes use rights, and understanding how these discourses and subject 'positionings' emerge also requires an appreciation of the wider socio-historical setting.

What happened during the early phases of water allocation reform in South Africa was that prevailing perceptions of scarcity and of managing water resources in a holistic way (in accordance with IWRM principles) dominated the policy process. Through the emphasis on scarcity and making efficient use of scarce resources, the notion of redistribution became hinged on the capacity to use water efficiently. Even if this may seem self-evident at first, it arises out of particular framings, which in turn point to equally self-evident solutions. Questioning the manner in which the problem is framed in the first place opens up the 'closure' achieved through advocating a particular solution (Fischer 2003; Stirling 2005).

Narrowing the frame of the problem: Naturalising scarcity

The narrow framing of the problem in the water allocation reform externalises past practices and presents scarcity as a natural phenomenon. But scarcity emerges as a consequence of the relations between means and ends and a social process that makes this relationship communicable, argues Toye (2005; see also Mehta 2010). The combination of value and scarcity in a particular society is not an objective fact, since values are socially determined. Historical legacies have, to a great extent, shaped water use patterns, through the massive land grabs in colonial times that were further reinforced through discriminatory legislation, and through the hydraulic mission embarked on by the State to provide farmers with ample irrigation water. The cheap supplies of water thus made available, together with the massive subsidies from the State, enabled the expansion of irrigated agriculture and saw the emergence of a featherbedded and highly inefficient agricultural sector (see Chapter 2: 16). Comparatively little attention has been paid to the scale of the resource capture enacted under apartheid:

> Resource capture is a social effect of environmental scarcity where more powerful groups of people manage to monopolise access to a critical resource such as water (Homer-Dixon 1994b: 11; Homer-Dixon & Percival 1996; Ohlsson 1998: 4; Ohlsson 1999: 38), thus leading to the ecological marginalisation of weaker groups of people. In short, resource capture, first of land and then of water, had been institutionalised by the mid-1960s, and all subsequent legislation was merely meant to tie up loose ends. This institutionalised resource capture became one of the cornerstones of apartheid strategy, but oddly enough is seldom written about by political scientists and analysts.
> (Turton & Henwood 2002: 50)

Chapter 3 described how, through privileged accounts, existing lawful uses were portrayed as beneficial and productive. Such accounts effectively obscure the fact that large-scale supply schemes serving mining and agricultural interests, and the huge subsidies lavished on the agricultural sector, contributed to wasteful and inefficient use. These accounts arose out of, and were nurtured by, the prevailing political economy conditions. In particular, the emphasis on the notion of 'two economies' served to portray the commercial sectors as the modern engine of economic growth, into which the 'traditional' sectors needed to be integrated (see Chapter 2: 28). The BEE policy further contributed to reinforcing the legitimacy of this view, in terms of providing blacks access to the benefits of economic growth occurring in the 'modern' sector. This was particularly true with respect to the mining sector (through the passing of the Mining Charter in 2002, see Chapter 2: 28). The ubiquitous notion of the 'two economies' and the need to integrate the 'traditional' sector into the 'commercial' one also explains much of the emphasis in water allocation reform on HDIs making 'commercial' use of water.

The notion of scarcity thus becomes an isolated idea without roots in historical usage, and the portrayal of existing users as being beneficial and efficient effectively removes them from constituting part of the problem – they are externalised. This shows how language not only reflects, but constructs meaning through the creation of individuals as subjects in the social formation, which serves to sustain relations of dominance (Thompson 1984). Though professing to address historical ills – the huge inequality in access to water – it does not take issue with the proximate cause of the problem, but only its symptoms. The water reform construes the problem in a way that externalises the resource capture. This defines the problem narrowly as one of scarcity and how to distribute scarce resources.

This particular framing of the problem, coupled with the nature of the 'negotiated settlement' (see Chapter 2: 24), which in effect preserved colonial land grabs by constitutional sanction (Hendricks & Ntsebeza 2000), provided the rules of formation for the discourse on existing users making beneficial and productive use. 'The prevailing political rhetoric, which placed an emphasis on the reallocation of land, water, and other natural resources to the people, was merely expedient during the decades of struggle for liberation and the early years of democratic transition' notes Francis (2005: 15), drawing on John Pilger.[2] The assertion that water legislation was non-discriminatory (as claimed by the author of the National Water Resources Strategy, see Chapter 3: 73) effectively discredited its evolution and the embedding of the British doctrine of riparian rights, which firmly linked access to water to land that was highly unequally distributed through the discriminatory Land Acts.

By explicitly advocating that existing users were unilaterally beneficial, the problem was framed merely and narrowly as one of 'distributing scarce resources equitably'. The reform process thus excluded past practices of land grabbing, pollution, waste and large-scale supply-oriented projects as the proximate causes of scarcity. Rather, scarcity was made out to be a natural condition, and coping with this productivity – understood as economic productivity – became a key principle. In essence, the discourse that emphasised the need to protect existing interests for the sake of the economy served to narrow the State's room for manoeuvre.

Equity hinging on efficiency

The idea of equitable access was linked with making productive use of water. Through a discursive strengthening of property rights (see Chapter 3), the State effectively entrenched existing interests. In order to accommodate the objective of equity, emphasis was shifted from redistribution to 'making the pie bigger', largely through the curbing of unlawful use. Having framed the problem in this manner, a path-dependence was created in which the pervasive notion was how to make the most efficient use of a scarce resource. 'Beneficial use in the public interest' is implicitly taken to mean economically productive use contributing to growth, made evident in the phrase '...provides social stability, which in turn promotes economic growth' (see Chapter 3: 62). Even social stability is not an end in itself: it is a means to an end, the end being greater economic growth. Thus, public interest is any use of water that contributes to economic growth. This indicates that the guiding criterion for policy is the greatest good for society, effectively a utilitarian view (Francis 2005: 15).

This way, equity became contingent on the ability to engage in efficient and productive use. This conditioning of equity upon the productive capacity of potential users takes an ahistorical view. It neglects the conditions through which many current water users gained access to their water resources, in particular through the government's large-scale efforts to solve the problem of 'poor-whiteism' following the Second World War (Sparks 2003b) through land allocation, the generous granting of water, and the subsidies lavished upon farmers (see Chapter 2: 16). No conditions of efficiency or sustainability were imposed on these users. On the contrary, they were at the receiving end of the State's hydraulic mission.

It also ignored the wider socio-economic context and the emphasis on competition and deregulation voiced in the embracing of the macroeconomic GEAR policy (see Chapter 2: 23). An emerging farmer embarking on 'the path to becoming commercial and competitive', as expressed in the Water Allocation Reform policy, would find this increasingly difficult in the current climate of deregulation of the agricultural sector, where even the protected sugar industry is facing greater competition (Lorentzen & Cartwright 2006). This detachment from macroeconomic policy conditions fails to acknowledge the challenges to be faced in terms of establishing a cadre of commercial black farmers or entrepreneurs. This has been repeatedly pointed out by scholars on land reform and agrarian change:

> One result of the deregulation and liberalisation process has been a growing rift in the sector between 'winners' and 'losers', with a rise in the rate of bankruptcies and the consolidation of landholdings into fewer hands…The state's support of an emerging class of black commercial farmers now sits uneasily with its removal of subsidies and other supports, which have combined to produce a uniquely hostile environment for new entrants into agriculture.
> (Hall 2004: 220)

The quote above clearly demonstrates the limited potential that exists for emerging farmers to transform their production activities into commercial enterprises, at least if the current macroeconomic conditions remain largely unchanged. But the political economy situation also produced a rather peculiar outcome. Intent on protecting property to shield 'the first economy' (see Chapter 2: 28) from a perceived threat of disruption, it paradoxically ended up protecting commercial agriculture (through the retention of existing lawful uses) which no-one but the agricultural lobby itself really desired. As observed in Chapter 3, neither of the discourses favoured the commercial agricultural sector, which had historically enjoyed a powerful political status. Although the institutionalist/industrialist perspective spurned commercial agriculture for its perceived marginal contributions to the economy, and the agriculturalist/livelihoods perspective concentrated on the notion of 'emerging' in terms of agriculture, viewing it more as a route to emancipation, the particular political-economic conditions prevailing at the time, which saw the protection of property rights as key to gaining investor confidence, meant that the agricultural lobby was indirectly favoured by the reform's insistence on retaining the notion of existing lawful use. This points to another feature, namely the tendency wherein issues of a political nature come to be 'bureaucratised' and how technical and 'self-evident' terms – such as 'existing lawful use' – come to cloud inherent political decision-making.

The discursive conceptions of justice in the 2006 water reform policy revolved around the idea of efficiency and productivity. By premising equitable access on productive capacity, the emerging policy narrative did two things. Firstly, it framed the problem mainly in terms of physical scarcity without paying attention to its proximate causes. Secondly, the key criterion of allocation was first and foremost utilitarian, rather than egalitarian, in that only those who would use water productively would be considered eligible.

The separation of land and water: Parallel processes, detached dynamics

Chapter 2 described how there was a resurgence of traditional authority following the transition, with traditional leaders gaining more powers through the passing of the Traditional Leadership and Governance Framework Act (No. 41 of 2003) (TLGFA) and the Communal Land Rights Act (No. 11 of 2004) (CLaRA). Chapter 3 described how the failure to link land reform with the water reform process resulted in scant attention being paid to the dynamics of land reform questions – it was largely a peripheral matter, rather than an integral part of the deliberations.

As Chapter 4 on water allocation in the Inkomati made clear, land is an important aspect of water allocation reform. It highlights how the diverging trajectories of access to land and water create new spaces for negotiation, and how access to land shapes people's access to water. This is made evident through the practice in which traditional leaders used development projects such as the sugar schemes to reward their allies (King 2005: own observations). Moreover, the alliances of sugar farmers and the struggles of small-scale farmers contrasted with the simplistic categorisations of users – or 'subject positionings' – in policy as 'ELUs' or 'HDIs'. In other words, the alliances between different user groups on the ground were much more complex and dynamic than what was reflected in the policy discourse.

The livelihoods discourse, though explicitly addressing equity and de-linking it from efficiency, is also culpable of ignoring power patterns and retaining an essentialist view that lacks nuance. The idea of the customary, supported by some of the adherents of the legal pluralism paradigm, does not take adequate account of the fact that customary law may also be regarded

as essentially a creation central to the colonial project of institutionalising difference, and its contents as determined in a dialogue between chiefs and administrators and both reflective and constitutive of the power relations at the time.
(Oomen 2005: 21)

The increasing demand for water in the Inkomati, emanating from the sugarcane schemes described in Chapter 4, was treated as a long overdue active demand from HDIs and lauded as a necessary trigger to speed up the pace of reform. However, scant attention was paid to issues of how land access governed water access, both in terms of there being little good land available, and in the processes of land allocation which were still largely couched in chiefly practices. Whilst it underscored the inadequacy of the discourse of productivity to account for local power relations, it also brought to light the assumptions inherent in the customary discourse, that did not pay sufficient attention to power issues within the context of traditional authorities, and how these work to constrain opportunities of access.

Severing the link between water and land opened up the way for the evolution of two different trajectories of access, where access to the one (water) is governed by the State, whilst land is still in many areas under chiefly authority, as made clear through the Inkomati case, and which was in the process of being reinforced through legislation such as the Communal Land Rights Act (Claassens & Cousins 2008). The CLaRA was declared unconstitutional on 11 May 2010.[3] Further, the process of settling land claims under the land reform process has a profound effect on the nature of water allocation reform and how it should be carried out.

Since the time of fieldwork, much has happened with respect to land reform in the Inkomati, and settling land claims has become the main vehicle of redistributing water, rather than through the water allocation reform itself. However, there are still considerable disparities in allocations, and the fact that water is redistributed through the settlement of claims only serves to highlight the importance of better co-ordination between the land and water departments (CMS 2009). After the DFID project came to an end, DWAF initiated a new project to see through the water allocation reform, and develop a Water Allocation Monitoring Index (WAMI), which was discontinued in 2009. Anderson, Mahlungu et al. (2008) drawing on existing data, estimate that more than 30 per cent of water has been reallocated to HDIs in the Inkomati basin. This had largely happened due to land reform, but this estimate includes claims that have yet to be settled. Also, there were instances where farmers sold off the water rights belonging to land under claim, rendering it virtually worthless. These trends further serve to underscore the need to co-ordinate the activities of the land and water departments and respective reform efforts.

The difficulty of determining use

The move from redistribution to finding more water came to emphasise the identification of unlawful use through the verification and validation process (see Chapter 4: 121), a process fraught with difficulty and overly optimistic trust in the ability of technology – the SAPWAT model and remote sensing – to ascertain rights. Taking into account the close connections between use rights, technology and science, there is a need to acknowledge that 'clear and unambiguous rights' cannot be 'proved' through technological/scientific approaches. This highlights the way in which rights are also fashioned through particular technologies. The technological determination of use rights occurs through monitoring, but also by means of technologies which control stream flow, such as those of dams and rivers, and by micro-monitoring, such as measuring flow rates and abstraction. However, the shift from the loosely-oriented common-property system of correlative riparian rights to the semi-individuated system of licences is difficult to sustain through the reliance on technologies for monitoring. In addition, the dynamics of land transactions have made the situation even more complex. At the time of writing, a new verification and validation project, expected to take some three years to complete, is underway.

A clean break with the past was thus made problematic in the early phases of the water allocation reform process, in part due to the discursive construction of rights, with the emphasis being put on use rights that contributed to the economy. Growth, rather than equity and redistribution, became the overriding issue for South African water reform, with environmental interests allowed to dictate a highly sophisticated and technical legislation (see Chapters 3 and 4), which points to the idea that the government did not take a strong enough stance with respect to redistribution. This perception aligns with similar views about the land reform process (see Chapter 4, also Hall 2007; Ntsebeza 2007; Ntsebeza & Hall 2007).

What of the future?

The status of WAR, and the emergence of new strategies

Policy is, as mentioned earlier, a moving target. The following section does not intend to offer an exhaustive review of recent developments, but rather point to some key events and trends. Things have moved on since the publication of the 2006 Water Allocation Reform (WAR) policy. In September 2008, a new, revised version of the WAR was completed. This was far more explicitly egalitarian, with specific targets set for redistribution. For instance, it stated that '30% of all water should be redistributed to Blacks by 2014, 45% by 2019 and 60% by 2024, and half of this should be targeted to women'. It says that targets are needed in order to measure the failure or success of the WAR programme. However, some estimates exist: according to the Department of Water Affairs' Strategic Plan 2009–2014, approximately 15 per cent of water licences have been allocated to HDIs. Although it is possible to some extent to trace the redistribution of water via the WARMS (Anderson, Mahlungu et al. 2008; De Jong 2010), one of the key problems with the redistributive efforts in this sense is that it relies mainly on the current heavily burdensome system of licensing. Moreover, the revised WAR policy has more or less been 'crowded out' by the new strategy of Water for Growth and Development (WfGD) that was launched in March 2009 by the then minister, Lindiwe Hendricks. The WfGD framework focuses on the idea of hydrological security (taking its cue from a 2006 World Bank paper by Grey and Sadoff (Grey & Sadoff 2006). The framework was developed over two years in consultation with key players in the water sector. A general theme is the notion that water is central to 'generate wealth, mitigate risk, and alleviate poverty'. The recommendations include placing water at the heart of all development decisions, 'exploring but not exploiting all sources of raw water to supplement the limited resources we now depend on'. This can be read in several ways – one way being the fact that there is more groundwater available than was originally presented (see Chapter 4: 89). It puts much emphasis on changing attitudes towards water, to save and conserve it. The emphasis is on growth, rather than broad-based development or transformation of the rural countryside.

Capacity constraints

Capacity is a core issue in terms of water allocation. To expand on that, and perhaps to put it glibly, in the effort to allocate and regulate water resources, the department has had to contend with capacity constraints, corruption and consultants. Relating to the issue of capacity at the departmental level, the troublesome recent history of the department (which was renamed the Department of Water Affairs[4] in May 2009) does not inspire confidence. It has been plagued by difficulties in attracting people with sufficient competence, and is suffering from high staff turnover rates. Moreover, it has to deal with allegations of corruption and infighting, as the following brief overview demonstrates. In September 2007, Jabulani Sindane, the Director-General (DG) who took over following Mike Muller's departure in 2005, quit. The department had been experiencing a deteriorating financial management situation and repeated adverse audit opinions from the auditor-general. The suddenness of Sindane's departure fuelled speculations that he had left as a result of infighting, and that the minister had made him the scapegoat for the department's financial troubles, as he was also unhappy with the department's track record on service delivery (*Mail & Guardian* 7 September 2007). The fact that as many as 52 of the

senior-level management staff had left since January that same year attests to the
level of frustration, so whether Sindane was made a scapegoat or whether he was
forced to resign due to incompetence remains open to conjecture. After Sindane
left, the department remained Director-less for six months, until Minister Lindiwe
Hendricks appointed Pam Yako as the new Director-General in March 2008. She,
too, resigned under suspicion of financial irregularities after only a few months
in office, but there were indications that a 'black caucus' within the department
had been intent on throwing her out because of her perceived 'harsh management
style' and her intention to clean up the department (*Mail & Guardian* 22 September
2009). The reform-minded Buyelwa Sonjica took over leadership of the ministry
for the second time in May 2009, but another cabinet reshuffle saw her post
being taken over by Edna Molewa in November 2010, amid fresh allegations of
financial irregularities committed by some of the top officials, including the acting
Director-General.[5] Such circumstances are not conducive to long-term capacity
building. Currently the department needs to hire consultants to compensate for lost
competence. Many of these have been former employees in the department, and
are now doing more or less the same job on a consultancy basis. This does raise the
question of how the department is being run, and how present and future capacity
needs will be met. Having said this, however, it is imperative to point out that there
are a number of deeply committed and professional individuals both inside and
outside of the department, regional offices and CMAs, who are working very hard to
improve the situation.

Challenges of regulation

The department had set itself the major task of facilitating a process of institutional
re-alignment. This project seeks to question the initial institutional framework,
and in particular flags the reduction and reconstitution of Catchment Management
Agencies (CMAs). The National Water Act presented the idea of a CMA as critical
for implementing IWRM effectively, and the National Water Resources Strategy
(NWRS) proposed that 19 CMAs be established, one for each Water Management
Area (WMA). However, progress has been extremely slow. In the past 10 years only
three CMAs have been established, and only two of these are fully operational (and
not even to those have full powers been devolved). A key problem has been that
the CMAs compete with the department for competent staff – a gain to the CMA
is a loss to the department. Moreover, the department is acting as proto-CMAs in
some areas whilst it is supposed to oversee and regulate the agencies, a situation
which creates a conflict of interests. The wisdom of hydrological boundaries is
increasingly being disputed (interview with Charles M'Marete, 27 July 2010; see also
Pollard & Du Toit 2005). This touches on a key issue that is a burgeoning field in

its own right, namely the nature of the representation of interests, or 'participatory management' (Anderson 2005). As was seen in Chapter 4, although the CMA had 14 representatives representing different water users' interests, often people 'on the ground' did not know who these representatives were, let alone how they would voice their concerns and interests. There seems to be a need to make better use of the institutions that do already exist, such as the district agricultural offices. However, there are projects underway focusing on the issue of enabling empowerment through the participatory management of redistribution processes,[6] and the recently approved Catchment Management Strategy for the Inkomati puts much store by engaging stakeholders. Still, Brown cautions against assuming too much with respect to participation (Brown 2011).

The department has to deal with these, and a whole set of other regulatory issues. At the time of writing, a discussion document has been published on how to craft regulation for the South African water sector. One of the most important points realised by this document, is the need to *simplify* (Karar 2010). What has become very clear is the burden of bureaucracy that a regulatory system of licences carries with it. The attempt to monitor every single licence holder means that the whole system becomes cumbersome and unwieldy, and does not reflect on-the-ground practices. The problems that small-scale farmers experience in gaining access to water, coupled with the long processing times of licence applications, encourages commercial users to go ahead and use water illegally, while marginal or potential water users are excluded, a point emphasised by Van Koppen (2006). Small-scale users need to be prioritised to a greater extent than has been the case hitherto through, for example, the use of General Authorisations. Attempts have been made to integrate the WARMS database with a licence application tracking system, the Water Use Licence Application Tracking System (WULATS), but this does not work very well (interview, M'Marete, 27 July 2010, see also De Jong 2010 for a detailed description). The department has had a hard time dealing with the backlog of licence applications, and in July 2010 they had a backlog of some 3 000 licences (interview, M'Marete, July 2010).

A particularly tricky issue has been dealing with the pollution from the mines and, as noted previously, AMD poses severe environmental problems. Rather than having the regional offices deal with the issues locally, a mobile team of the most competent and experienced people was put together, to deal with pending licence applications and assess their environmental implications. Following the logic of simplification, the important thing is not to deal with each and every water user, but to regulate the large-scale users who potentially have the greatest impacts. Future prospects hinge on the department's ability to find a way out of its current disarray, and to develop a way of regulating the use of water resources that is better adapted to the available skills and capacity.

The politics of redistribution

Arguing that water resource allocation and regulation needs capacity and an adequate regulatory system is necessary, but even more necessary is an appropriate overall political situation within which such regulation takes place. Making the case for redistribution implies the need to think about possible scenarios. The greatest challenge is the lack of political will to come up with a comprehensive strategy for what sort of path should be pursued to facilitate rural change and agrarian transformation. In a recent article in the magazine *Amandla*, Cousins and Hall (2010) point to the fact that little is happening with respect to the creation of a new comprehensive rural development strategy. A Green Paper on land reform and rural development has been tabled, but at the time of writing had yet to be put forward. The current suggestions are vacuous and involve merely 'tinkering' with market mechanisms, according to Cousins and Hall.

Popular perceptions that radical changes were afoot with the inauguration of a new government in 2009 are fading. Thabo Mbeki's resignation from the office of President in 2008 paved the way to power for Jacob Zuma, who enjoyed wide popular support, despite being a highly controversial political figure (Southall & Daniel 2009). Zuma has found himself mired in a series of controversies, and the uncertainty and infighting that has beset the ANC for several years will most likely not go away anytime soon. Although Zuma still enjoys wide support among the grassroots of the ANC, who were increasingly alienated by Mbeki's aloofness, the charges of corruption and rape that have been levied against him mean that his aptitude for the highest office of the country is doubted by many. This creates a climate of uncertainty and apprehension about the country's future. Zuma was brought to power thanks to the breadth of the ANC coalition, but this also renders him somewhat hamstrung in terms of taking clear positions with respect to policies: he seems to have adopted a consensual and pragmatic policy-making style as his trademark. His integrity questioned, he has also been accused of political weakness (among other things, for failing to put the rancorous ANC Youth League leader, Julius Malema, in his place). There are a number of challenges facing Zuma, who is vulnerable to criticism that reform has been far too slow. There is a need for a strong political force and a clear vision of where the country will be going in the near future. Zuma prides himself on having achieved more effective governance with the restructuring of the government and the reconfiguration of ministries, but the reshuffling of ministries signals a lack of attention to the challenges facing the countryside, and the need to integrate land and water resources more effectively in order to reduce poverty and provide effective implementation routes for the redistribution of resources.

Despite the fact that rural development figures were one of the five key themes of the ANC's election manifesto (ANC 2009), there are indications that what is happening on paper is being countered by what is happening in practice. In terms of water and land rights, the experience of the former homeland areas in the Inkomati (see Chapter 4) suggests that the different trajectories of access to land and water create new spaces for negotiation that need to be taken into account. Issues of marginal land quality and traditional structures of governance of land access are key in determining water access. Researchers on land reform issues have long battled with these problems, and there is a strong need for the water reform processes to link to these debates, and to explore further what a livelihoods approach would mean in practice. The failure to link the different dynamics of the two routes of access holds the risk of ignoring the power patterns that may place reform efforts in further peril. The ANC's 2009 election manifesto does acknowledge the need to better link ongoing water and land reform: 'Ensur[ing] a much stronger link between land and agrarian reform programmes and water resource allocation and ensur[ing] that the best quality of water resources reach all our people, especially the poor' (ANC 2009: 9). However, a major overhaul of the ministries has resulted in the creation of a new Ministry for Rural Development and Land Reform (MRDLR) and a Ministry for Agriculture, Forestry and Fisheries (MAFF), which sits uneasily with the promise to ensure a stronger link between land and agrarian reform programmes (ANC 2009). What this restructuring effectively amounts to is a perpetuation of the dual economy thinking, through creating a 'ministry for the rich' and a 'ministry for the poor' (Hall 2009: 6). There are some positive signs of change, however, in particular the coming together of researchers, consultants and policy-makers to address the structural challenges, in which land and water figure as a key focus (Schreiner, Tapela et al. 2010). The hope is that these efforts will effectively permeate policy-making, and that water allocation is driven by a truly development-oriented agenda.

The important thing, at the end of the day, is to see water reform as one of the key elements in transforming South Africa into a less glaringly unequal society, and to reduce the current gap between the haves and the have-nots. This requires concerted and focused efforts in a range of sectors, in which water is one of the vital issues. What prospects exist for real redistributive efforts to emerge? South Africa still faces profound challenges in terms of high levels of inequality and poverty, as the following quote testifies:

> South Africa is rich by African standards, but has nevertheless a very high poverty rate – of more than 34 per cent in 2000. The total poverty gap is less than one per cent (the production tax) of GNI. The huge inequalities of the

country are inherited from apartheid. But since the ANC took over in the early 1990s, South Africa could have eliminated all its extreme poverty by a comparatively small tax on the non-poor, of just above one per cent in 2000. Having not done so can be interpreted as a sign that the process of social and political conciliation after the war has also led to miserly behaviour towards the poor.
(Lind & Moene 2008: 13)

This 'miserly attitude', was quite strongly reflected in the original WAR, through its emphasis on the entrenchment of existing water users and through the linking of equity with productivity. In the new strategy, where specific goals have been identified, equality in resource distribution is a far more unequivocal objective. Whether these objectives will be achieved within the stipulated time frames, given the difficulties experienced so far, will to a large degree depend on the presence of political will. There is a strong need to carve out a coherent development strategy that takes an integrated view of land and water issues and rural development. A concerted effort to map out the dynamics of water and land access and identify viable pathways for improving people's lives in the rural areas is needed. Some work is already being done, through, for instance, studying the role of water in informal economies.[7]

Wider implications

Although South Africa represents a unique situation, the lessons to be drawn from its experiences have wider implications for water rights reform elsewhere. The tensions inherent in the IWRM paradigm, between central oversight and decentralised management, are not confined to the South African context. Bringing water resources wholly under the State and departing from approaches that derive from a legal principle, such as the prior appropriation or the riparian doctrine, will generally provide governments with discretionary power and latitude for different interpretations of how use rights should be understood, defined and allocated. As has been argued throughout this book, the way use rights are understood and allocated hinges in part on the nature of the narratives that emerge during the policy process, and how these narratives portray existing and potential users. Thus, use rights are discursively produced – how rights and potential rights holders are talked about will be constitutive of the nature of rights themselves, as well as the pattern of allocation.

Although the allocation of water resources in the South African context was initially explicitly political – with the aim of bringing about a more equitable distribution – the process of allocation reform ended up couching inherently political issues in innocuous-sounding and neutralising terms such as existing lawful use. With the recent identification of measurable goals of redistribution in the recently launched Water Allocation Reform Strategy, politics has been firmly returned to the policy, though its realisation will be subject to the presence of political determination to see that it is placed within a broader strategy of agrarian transformation. The process of depoliticising policy is a risk in another setting as well: that questions of water allocation that were previously founded on more or less easy-to-define legal principles are now subject to State deliberations through the issuing of State-authorised licences or permits. Though seemingly objective criteria are deployed to aid decision-making in favour or disfavour of any given individual applicant or group for a water-use right, these criteria are liable to be deeply influenced by the particular political economy context and the associated discourses that emerge. Over time, this influence is likely to be camouflaged in technical or legal language, and may mask inherently political decisions as purely technical or bureaucratic ones. It is, therefore, necessary to be aware of the way in which the policy process can contribute to obscuring the power constellations present in particular narratives around rights and rights holders.

In addition, by giving the State the authority to issue use rights, the onus of defining the content of such rights is also placed on the State. In situations in which government capacity is stretched to the limits, as is the case in many developing countries, this may be an almost insurmountable task and a questionable use of State capacity. But even in developed country settings the process of attempting to accurately quantify the portion of a resource to which a user holds rights is fraught with problems. One issue concerns the sheer indeterminacy inherent in the nature of water flows. Even the best hydrological models can do no more than provide approximate estimates of flows and hydrological patterns, as Jasanoff (2004) has pointed out. The outputs of models are only as good as the guesswork that went in, in terms of deciding on parameters and variables. The tendency towards rigid determination of use rights is undermined by the presence of incertitude, which further renders the task of backing legal claims with reference to scientific quantification moot. The difficulties in terms of quantifying and legally backing up State-authorised use rights seriously bring into question their merit. Finding ways to deal with hydrological incertitude would likely provide a more fruitful pathway to managing water resources than the current pattern of attempting accurate quantifications.

These observations open up broader questions about State–citizen relations, and broader goals of resource distribution and development. Implementing IWRM and institutionalising use rights ultimately concerns issues of State versus citizen authority over resources, and how that authority is negotiated. This insight implies that attempts to implement IWRM and associated permits or licensing systems in different contexts needs to be explicit about the political nature of such reform efforts, rather than treat them as purely technical or management exercises. There is, therefore, a need to raise awareness in terms of the ways in which discourses arise, and how language at the policy level is deployed in such a manner that it contributes to fashioning property relations, and to make sure that the inherently political nature of defining and allocating use rights does not become obscured.

Notes

1. Which may be capped through the use of permits, see Chapter 2: 42.
2. John Pilger works as a journalist for the British newspaper, *The Guardian*.
3. See http://www.lrc.org.za/press-releases/1227-2010-05-11-communal-land-right-act-declared-unconstitutional
4. The Forestry portfolio has been integrated into the Ministry of Agriculture, Forestry and Fisheries (MAFF).
5. http://cer.org.za/?p=1410
6. See e.g. http://www.pegasys.co.za/what-we-do/water-strategy/#1
7. See http://ongoing-research.cgiar.org/factsheets/water-rights-in-informal-rural-economies-in-the-limpopo-and-volta-basins/

Appendix A

Note: Dates of interviews are given in the text

Interviewees at policy level

Department of Water Affairs and Forestry	Chief Co-ordinating Officer (COO)	14
	Director, Water Allocations	
	Deputy Director, Water Allocations	
	Director, Water Abstraction and Instream Use	
	Senior Adviser, Water Abstraction and Instream Use	
	Director, Regional Co-ordination and Support	
	Deputy Director, Stream Flow Reduction and Strategic Environmental Assessment	
	Senior Adviser, Stream Flow Reduction and Strategic Environmental Assessment	
	Acting Director, Legal Services	
	Director, Water Resources Planning	
	Director, Policy and Strategy Co-ordination	
	Senior Adviser, Water Resources Planning	
	Adviser, Water Resources Planning	
	Representative, Joint Water Commission (JWC)	
Department for International Development, UK (DFID)	Consultant	1
Water Research Commission	Director, Water Resources Management	1
CSIR	Senior Researcher (and founder of AWIRU)	1
University of Pretoria	Sylvain Perret, Associate Professor, Department of Agricultural Economics	2
	Stefano Farolfi, environmental economist	
Lawyers (members of Water Law drafting team)	Hadley Kavin	2
	Francois Junod	
Clear Pure Water	Andrew Pott	2
	Jason Hallowes	
Independent consultants	Marna de Lange, Socio-technical Interfacing	2
	Cartwright Consultancy	
Water for Food Movement	MaTshepo Khumbane	1
PLAAS	Ben Cousins, Edward Lahiff, Barbara Tapela	3
International Water Management Institute (IWMI)	IWMI director, IWMI researchers	4

Other academics	Christo de Coning, University of Witwatersrand Vuysile Zenani, UNDP	2
No. of interviewees, policy level:		**35**

Interviewees at regional/local level

Department of Water Affairs and Forestry – Regional Office	Deputy Director Two senior staff	3
Provincial Department of Agriculture and Land Affairs (DALA)	Head of Technical Support Services Mampho Malgas, Head of Restitution Committee Marc Anthony, CASP-Mpumalanga project leader	3
The Mpumalanga African Farmers' Union (MAFU)	Motsepe Matlala, President	1
Lowveld Agricultural College	Dr Reid, Senior Lecturer Marius van Rooyen	2
KOBWA	Enoch Dhlamini Water Bailiff	2
Inkomati Catchment Management Agency	Chief Executive Officer, Sizile Ndlovu Governing Board representatives	11
Nkomazi Municipality	Christo Smit, Chief Engineer, Water Services Head, Planning Division	2
MBB, consultancy firm	Koos van Rensburg	1
Komati River Irrigation Board	Chairperson	1
Lomati River Irrigation Board	Chairperson Engineer	2
Upper Komati Water Users Association	Chairperson	1
Elands River Irrigation Board	Hubert Neumann	1
First National Bank, Nelspruit	Stefan Scheepers	2
Land Bank, Nelspruit	Employee	1
Local lending business, Malelane	Employee	1
Sappi	Andre van Tonder, Consultant, environmental accounting	1
Nkomati Mine	Mr W Oosthuizen, Project Manager	1
Mpumalanga Cane Growers' Association	Brian Sugden, Director Lizana Stark, Secretary Hans Maphanga, Liaison Officer, Nkomazi East Vincent Quabe, Liaison Officer, Nkomazi West	4

TSB Sugar	Martin Slabbert	1
South African Sugar Association (SASA)	Banie Swart, Extension Officer (SASEX)	1
District Agricultural Offices	Philemon Mthembu, Chief, Technical Services Samuel Nyoni	2
Ndwandwa Trust	Steve Sikhosane, Director	1
Non-governmental organisations	Sharon Pollard, Derick du Toit, AWARD Philip Owen, GeaSphere	3
White commercial farmers	Mr Brown (Elands River) Mr Chance (Komati River) Ms Engels (Lomati River) Mr Gordon (Lomati River) Mr Johnson (Lomati River) Mr Kruger (Lomati River) Mr Middle (Komati River) Mr Noordwyck (Komati River) Mr Schoeman (Komati River) Mr Terence (Komati River) Mr van Dyk (Komati River) Mr Wasserman (Komati River)	12
Emerging sugar farmers	Mr Bila (Figtree C) Ms Khoza (Madadeni) Mr Khumbisa (Figtree A) Ms Lubisi (Figtree A) Ms Makenathe (Madadeni) Mr Mandlazi (Figtree A) Ms Mashele (Mangane) Mr Mashiye (Figtree C) Ms Mavundlu (Mfumfane B) Mr Mhlongo (Langeloop 1/Mabodweni) Mr Mkatshwa (Sikhlawane) Mr Mkhabela (Figtree B) Mr Mupisa (Mfumfane B) Mr Nkalanga (Spoons B) Mr Ndlovu (Madadeni) Mr Ndluli (Ngogolo) Mr Shongwe (Sikhlawane) Mr T Sibiya (Mfumfane A) Magudu farmers Sikwhalane farmers Tonga farmers	20+

Vegetable farmers	Khutselani Women's Group (Driekoppies)	1 (20+
Dryland farmers	Mr Mabaso (Chairperson, Steenbok Youth Farmers Association, growing cassava)	9+
	Ms Mahalela (cotton)	
	Mr Makushe (cotton)	
	Mr Mathebula (Chairperson, Siyavuma Cassava Farmers' Association)	
	Mr Sibitane (cotton)	
	Magodwane farmers' project	
	Siyanka cassava farmers	
Commercial farmers	Mr A Sibiya	
	Abundant Life Skills	2
Other	Legal practitioner based in Komatipoort	1
No. of interviewees, regional/local level		**93+**

Appendix B

Fundamental principles and objectives for a new water law in South Africa

Source: Government of South Africa (1997). White Paper on Water on Policy. Pretoria: Department of Water Affairs and Forestry

Legal aspects of water

Principle 1

> The water law shall be subject to and consistent with the Constitution in all matters including the determination of the public interest and the rights and obligations of all parties, public and private, with regards to water. While taking cognisance of existing uses, the water law will actively promote the values enshrined in the Bill of Rights.

Principle 2

> All water, wherever it occurs in the water cycle, is a resource common to all, the use of which shall be subject to national control. All water shall have a consistent status in law, irrespective of where it occurs.

Principle 3

> There shall be no ownership of water but only a right (for environmental and basic human needs) or an authorisation for its use. Any authorisation to use water in terms of the water law shall not be in perpetuity.

Principle 4

> The location of the water resource in relation to land shall not in itself confer preferential rights to usage. The riparian principle shall not apply.

The water cycle

Principle 5

> In a relatively arid country such as South Africa, it is necessary to recognise the unity of the water cycle and the interdependence of its elements, where evaporation, clouds and rainfall are linked to groundwater, rivers, lakes, wetlands and the sea, and where the basic hydrological unit is the catchment.

Principle 6

> The variable, uneven and unpredictable distribution of water in the water cycle should be acknowledged.

Water resource management priorities
Principle 7
> The objective of managing the quantity, quality and reliability of the nation's water resources is to achieve optimum, long term, environmentally sustainable social and economic benefit for society from their use.

Principle 8
> The water required to ensure that all people have access to sufficient water shall be reserved.

Principle 9
> The quantity, quality and reliability of water required to maintain the ecological functions on which humans depend shall be reserved so that the human use of water does not individually or cumulatively compromise the long term sustainability of aquatic and associated ecosystems.

Principle 10
> The water required to meet the basic human needs referred to in Principle 8 and the needs of the environment shall be identified as 'The Reserve' and shall enjoy priority of use by right. The use of water for all other purposes shall be subject to authorisation.

Principle 11
> International water resources, specifically shared river systems, shall be managed in a manner that optimises the benefits for all parties in a spirit of mutual co-operation. Allocations agreed for downstream countries shall be respected.

Water resource management approaches
Principle 12
> The National Government is the custodian of the nation's water resources, as an indivisible national asset. Guided by its duty to promote the public trust, the National Government has ultimate responsibility for, and authority over, water resource management, the equitable allocation and usage of water and the transfer of water between catchments and international water matters.

Principle 13
> As custodian of the nation's water resources, the National Government shall ensure that the development, apportionment, management and use of those resources is carried out using the criteria of public interest, sustainability, equity and efficiency of use in a manner which reflects its public trust obligations and the value of

water to society while ensuring that basic domestic needs, the requirements of the environment and international obligations are met.

Principle 14

Water resources shall be developed, apportioned and managed in such a manner as to enable all user sectors to gain equitable access to the desired quantity, quality and reliability of water. Conservation and other measures to manage demand shall be actively promoted as a preferred option to achieve these objectives.

Principle 15

Water quality and quantity are interdependent and shall be managed in an integrated manner, which is consistent with broader environmental management approaches.

Principle 16

Water quality management options shall include the use of economic incentives and penalties to reduce pollution, and the possibility of irretrievable environmental degradation as a result of pollution shall be prevented.

Principle 17

Water resource development and supply activities shall be managed in a manner which is consistent with the broader national approaches to environmental management.

Principle 18

Since many land uses have a significant impact upon the water cycle, the regulation of land use shall, where appropriate, be used as an instrument to manage water resources within the broader integrated framework of land use management.

Principle 19

Any authorisation to use water shall be given in a timely fashion and in a manner which is clear, secure and predictable in respect of the assurance of availability, extent and duration of use. The purpose for which the water may be used shall not arbitrarily be restricted.

Principle 20

The conditions upon which authorisation is granted to use water shall take into consideration the investment made by the user in developing infrastructure to be able to use the water.

Principle 21
>The development and management of water resources shall be carried out in a manner which limits to an acceptable minimum the danger to life and property due to natural or manmade disasters.

Water institutions
Principle 22
>The institutional framework for water management shall as far as possible be simple, pragmatic and understandable. It shall be self-driven and minimise the necessity for State intervention. Administrative decisions shall be subject to appeal.

Principle 23
>Responsibility for the development, apportionment and management of available water resources shall, where possible and appropriate, be delegated to a catchment or regional level in such a manner as to enable interested parties to participate.

Principle 24
>Beneficiaries of the water management system shall contribute to the cost of its establishment and maintenance on an equitable basis.

Water services
Principle 25
>The right of all citizens to have access to basic water services (the provision of potable water supply and the removal and disposal of human excreta and waste water) necessary to afford them a healthy environment on an equitable and economically and environmentally sustainable basis shall be supported.

Principle 26
>Water services shall be regulated in a manner which is consistent with and supportive of the aims and approaches of the broader local government framework.

Principle 27
>While the provision of water services is an activity distinct from the development and management of water resources, water services shall be provided in a manner consistent with the goals of water resource management.

Principle 28
>Where water services are provided in a monopoly situation, the interests of the individual consumer and the wider public must be protected and the broad goals of public policy promoted.

Appendix C

Sections 25, 27 and 35 of the National Water Act (No. 36 of 1988)

Chapter 4, part 1, section 25
Transfer of water use authorisations

(1) A water management institution may, at the request of a person authorised to use water for irrigation under this Act, allow that person on a temporary basis and on such conditions as the water management institution may determine, to use some or all of that water for a different purpose, or to allow the use of some or all of that water on another property in the same vicinity for the same or a similar purpose.

(2) A person holding an entitlement to use water from a water resource in respect of any land may surrender that entitlement or part of that entitlement –
  (a) in order to facilitate a particular licence application under section 41 for the use of water from the same resource in respect of other land; and
  (b) on condition that the surrender only becomes effective if and when such application is granted.

(3) The annual report of a water management institution or a responsible authority, as the case may be, must, in addition to any other information required under this Act, contain details in respect of every permission granted under subsection (1) or every application granted under subsection (2).

Chapter 4, part 2, section 27
Considerations for issue of general authorisations and licences

(1) In issuing a general authorisation or licence a responsible authority must take into account all relevant factors, including –
  (a) existing lawful water uses;
  (b) the need to redress the results of past racial and gender discrimination;
  (c) efficient and beneficial use of water in the public interest;
  (d) the socio-economic impact
    (i) of the water use or uses if authorised;
    (ii) of the failure to authorise the water use or uses;
  (e) any catchment management strategy applicable to the relevant water resource;
  (f) the likely effect of the water use to be authorised on the water resource and on other water users;
  (g) the class and the resource quality objectives of the water resource;
  (h) investments already made and to be made by the water user in respect of the water use in question;

- (i) the strategic importance of the water use to be authorised;
- (j) the quality of water in the water resource which may be required for the Reserve and for meeting international obligations; and
- (k) the probable duration of any undertaking for which a water use is to be authorised.

(2) A responsible authority may not issue a licence to itself without the written approval of the Minister.

Chapter 4, part 3, section 35
Verification of existing water uses

(1) The responsible authority may, in order to verify the lawfulness or extent of an existing water use, by written notice require any person claiming an entitlement to that water use to apply for a verification of that use.

(2) A notice under subsection (1) must –
- (a) have a suitable application form annexed to it;
- (b) specify a date before which the application must be submitted;
- (c) inform the person concerned that any entitlement to continue with the water use may lapse if an application is not made on or before the specified date; and
- (d) be delivered personally or sent by registered mail to the person concerned.

(3) A responsible authority –
- (a) may require the applicant, at the applicant's expense, to obtain and provide it with other information, in addition to the information contained in the application;
- (b) may conduct its own investigation into the veracity and the lawfulness of the water use in question;
- (c) may invite written comments from any person who has an interest in the matter; and
- (d) must afford the applicant an opportunity to make representations on any aspect of the application.

(4) A responsible authority may determine the extent and lawfulness of a water use pursuant to an application under this section, and such determination limits the extent of any existing lawful water use contemplated in section 32(1).

(5) No person who has been required to apply for verification under subsection (1) in respect of an existing lawful water use may exercise that water use –
- (a) after the closing date specified in the notice, if that person has not applied for verification; or
- (b) after the verification application has been refused, if that person applied for verification.

(6) A responsible authority may, for good reason, condone a late application and charge a reasonable additional fee for processing the late application.

References

Adger NW, Benjaminsen TA, Brown K & Svarstad H (2001) Advancing a political ecology of global environmental discourses. *Development and Change* 32(4): 687–715

Aguilera-Klink F, Pérez-Moriana E & Sánchez-García J (2000) The social construction of scarcity. The case of water in Tenerife (Canary Islands). *Ecological Economics* 34: 233–245

Alexander N (2006) Racial identity, citizenship and nation building in post-apartheid South Africa. Edited version of a lecture delivered at the East London Campus, University of Fort Hare

Aliber M & Mokoena R (2002) *The interaction between the land redistribution programme and the land market in South Africa: A perspective on the willing-buyer/willing-seller approach.* Land Reform and Agrarian Change in Southern Africa Occasional Paper No. 21, University of the Western Cape

ANC (2009) ANC Election Manifesto 2009 'Working together we can do more'

Anderson AJ (2005) Engaging disadvantaged communities: Lessons from the Inkomati CMA establishment process. Paper presented at the international workshop on African Water Laws: Plural Legislative Frameworks for Rural Water Management in Africa, Johannesburg, South Africa (26–28 January)

Anderson AJ, Mahlungu MS, Cullis J & Swartz S (2008) Integrated monitoring of water allocation reform in South Africa. *Water SA* 34(6): 731-737

Angelsen A & Vainio M (1998). *Poverty and the environment.* Bergen, Norway: Comparative Research on Poverty (CROP) Publications

Armitage RM (1999) *An economic analysis of surface irrigation water rights transfers in selected areas of South Africa.* WRC Report No. 870/1/99. Pretoria: Water Research Commission

Ashley C, Chaumba J, Cousins B, Lahiff E, Matsimbe Z, Mehta L, Mokgope K, Mombeshora S, Mtisi S, Nhantumbo I, Nicol A, Norfolk S, Ntshona Z, Pereira J, Scoones I, Seshia S, Wolmer W & Nyamu-Musembi C (2003). Rights talk and rights practice: Challenges for southern Africa. *Institute of Development Studies (University of Sussex) Bulletin* 34(3): 97–111

Backeberg GR (1995) Towards a market in water rights: A pragmatic institutional economic approach. Discussion Paper. Pretoria: University of Pretoria

Backeberg GR (2005) Water institutional reforms in South Africa. *Water Policy* 7: 107–123

Bate R & Tren R (2002) *The cost of free water: The global problem of water misallocation and the case of South Africa.* Johannesburg: The Free Market Foundation

Bauer CJ (1997) Bringing water markets down to earth. The political economy of water rights in Chile, 1976–95. *World Development* 25(5): 639–656

Benjaminsen TA & Lund C (Eds) (2003). *Securing land rights in Africa.* London: Frank Cass

Berkhout F, Leach M & Scoones I (2003). *Negotiating environmental change.* Cheltenham: Edward Elgar

Bond P (2000) *Elite transition: From apartheid to neoliberalism in South Africa.* London: Pluto Press

Bond P (2004) *Talk left, walk right: South Africa's frustrated global reforms.* Pietermaritzburg: University of KwaZulu-Natal Press

Bond P (2007) Transcending two economies? Renewed debates in South African political economy. *Special Issue of AFRICANUS, Journal of Development Studies* 37(2)

Bond P & Khoza M (1999) An RDP Policy Audit. Pretoria: Human Sciences Research Council

Boroto RJ (2004) A collaborative effort towards implementing IWRM: A southern African perspective. Paper presented at the International Conference on IWRM, Kyoto (6-8 December)

Bourdieu P & Thompson JB (1991). *Language and symbolic power*. Cambridge: Polity Press

Bromley DW (2004) Property rights: Locke, Kant, Peirce and the logic of volitional pragmatism. In: HM Jacobs. *Private Property in the 21st century: The future of an American ideal*. Cheltenham: Edward Elgar: 19–30

Brown J (2011) Assuming too much? Participatory water resource governance in South Africa. *The Geographical Journal* 177(2): 171-185

Brown J (2005) A cautionary tale: A review of the institutional principle in practice, with reference to the Inkomati WMA, South Africa. Manchester: University of Manchester

Brown J & Woodhouse P (2004). *Pioneering redistributive regulatory reform. A study of implementation of a Catchment Management Agency for the Inkomati Water Management Area, South Africa*. Centre on Regulation and Competition Working Paper Series No. 89. Manchester. Institute for Development Policy and Management

Bruns BR & Meinzen-Dick RS (Eds) (2000) *Negotiating water rights*. London: ITDG Publications

Bundy C (1988) *The rise and fall of the South African peasantry*. London: James Currey

Burchi S (2005) The Interface between customary and statutory rights. A statutory perspective paper presented at the international workshop on African Water Laws: Plural Legislative Frameworks for Rural Water Management in Africa, Johannesburg, South Africa (26–28 January)

Burger A (2006) A study of Roman water law with specific reference to water allocations and prior appropriation (amended and abbreviated by Dr Heather McKay). WRC Report No TT 279/06, South Africa Water Research Commission

Cairncross S (2003) Water supply and sanitation: Some misconceptions. *Tropical Medicine and International Health* 8(3): 193–19

Carmody P (2002) Between globalisation and (post) apartheid: The political economy of restructuring in South Africa. *Journal of Southern African Studies* 28(2): 255–275

Chereni A (2007) The problem of institutional fit in integrated water resources management: A case of Zimbabwe's Mazowe catchment. *Physics and Chemistry of the Earth* 32: 1246–1256

Chhotray V & Woodhouse P (2005) Legal and extra-legal channels for the redress of historical injustice: Water rights in India and South Africa. Paper presented at the International Workshop on Water Poverty and Social Crisis: Perspectives for Research and Action, Agadir, Morocco. Available https://ueaeprints.uea.ac.uk/33642/

Chikozho C & Latham J (2005). Shona customary practices in the context of water sector reforms in Zimbabwe. Paper presented at the international workshop on African Water Laws: Plural Legislative Frameworks for Rural Water Management in Africa Johannesburg, South Africa (26–28 January)

Cliffe L (2004) From 'African Renaissance' to re-empowering chiefs. *Review of African Political Economy* 31(100): 354–356

Claassens A (2003) *Community views on the Communal Land Rights Bill*. Research Report No. 15. Cape Town: Programme for Land and Agrarian Studies

Claassens A (2005) *The Communal Land Rights Act and women: Does the Act remedy or entrench discrimination and the distortion of the customary? Land reform and agrarian change in Southern Africa.* Occasional Paper No. 28. Cape Town: Programme for Land and Agrarian Studies

Claassens A & Cousins (2008) *Land, power & custom: Controversies generated by South Africa's Communal Land Rights Act.* Cape Town: University of Cape Town Press

Cobb R & Elder (1979) *Participation in American politics: The dynamics of agenda-building.* Baltimore: Johns Hopkins University Press

Commission on Restitution of Land Rights (2007) Massive land handover for Mpumalanga communities. Commission press release, 15 June 2007

Conca K (2006) *Governing water: Contentious transnational politics and global institution building.* Cambridge, MA: MIT Press

Cousins B (2005) Agrarian reform and the 'two economies': Transforming South Africa's countryside. In: R Hall & L Ntsebeza. *The land question in South Africa: The challenge of transformation and redistribution.* Athens, OH: Ohio University Press

Cousins B & Hall R (2010) Plotting a new course for land reform and rural development. *Amandla.* September/October 2010

Cullis J & Van Koppen B (2007) *Applying the Gini Coefficient to measure inequality of water use in the Olifants River Water Management Area, South Africa.* IWMI Research Report 113. Colombo: International Water Management Institute

De Coning C (2006) Overview of the water policy process in South Africa. *Water Policy* 8: 505–528

De Jong F (2010) *Water allocation reform though licensing: The effect of neoliberalism on access to water for Historically Disadvantaged Individuals, Limpopo Province, South Africa.* Law and Governance & Irrigation and Water Engineering Group. Wageningen: Wageningen University MSc: 138

Denzin NK & Lincoln YS (Eds) (2003) *Collecting and interpreting quantitative materials.* London: Sage Publications

Department of Water Affairs and Forestry (1999). A pricing strategy for raw water use charges. Pretoria: Department of Water Affairs and Forestry: 1353

Department of Water Affairs and Forestry (2004a). National water resources strategy. 1st edition. Pretoria, Department of Water Affairs and Forestry

Department of Water Affairs and Forestry (2004b). WMA05 Inkomati: Internal strategic perspective. Department for Water Affairs and Forestry. Pretoria: Department of Water Affairs and Forestry

Dhlamini E, Dhlamini S & Mthimkhulu S (2007) Fractional water allocation and reservoir capacity sharing concepts: An adaptation for the Komati Basin. *Physics and Chemistry of the Earth* 32: 1275–1284

Du Toit A & Neves D (2007) In search of South Africa's 'Second Economy'. *Africanus* 37(2): 145–174

Duraiappah AK (1998) Poverty and environmental degradation: A review and analysis of the nexus. *World Development* 26(12): 2169–2179

Eales K (2011) Water Services South Africa 1994-2009. In B Schreiner and R Hassan (Eds) *Transforming water management in South Africa: Designing and implementing a new policy framework.* Dordrecht: Springer

Eggertsson T (1996) The economics of control and the cost of property rights. In: S Hanna, C Folke & KG Mäler. *Rights to nature: Ecological, economic, cultural, and political principles of institutions for the environment.* Washington DC: Island Press: 157–179

Fairclough N (1995) *Critical discourse analysis: The critical study of language.* Harlow: Longman

Falkenmark M (1998) Dilemma when entering 21st century: Rapid change, but lack of sense of urgency. *Water Policy* 1: 421–436

Falkenmark M & Widstrand C (1992) Population and water resources: A detailed balance. *Population Bulletin* 47(3): 1–36

Faysse N & Gumbo J (2004) The transformation of irrigation boards into water user associations in South Africa: Case studies of the Umlaas, Komati, Lomati and Hereford Irrigation Boards. Working Paper No. 73, part 2. Colombo, Sri Lanka: International Water Management Institute.

Fine B (2003) Review: The political economy of transition in South Africa. *Journal of Southern African Studies* 29(2): 571–573

Fischer F (2003) *Reframing public policy: Discursive politics and deliberative practices.* Oxford: Oxford University Press

Fischer F & Forester J (Eds) (1993) *The argumentative turn in policy analysis and planning.* Durham and London; Duke University Press

Forsyth T (2003) *Critical political ecology: The politics of environmental science.* London: Routledge

Foucault M (1980) *Power/knowledge: Selected interviews and other writings 1972–1977* Brighton: Harvester Press

Foucault M (1991) Politics and the study of discourse. In: G Burchell, C Gordon & P Miller. *The Foucault effect: Studies in governmentality.* Chicago: The University of Chicago Press: 53–73

Fourie C (2001) Land and property registration at the crossroads. *Habitat Debate* 7(3): 16

Francis R (2005) *Water justice in South Africa: Natural resource policy at the intersection of human rights, economics, and political power.* Expresso Preprint Series Working Paper No. 518. The Berkeley Electronic Press

Fraser N (1989) Talking about needs: Interpretive contests as political conflicts in welfare-state societies. *Ethics* 99: 291–313

Freudenburg WR (2005) Privileged access, privileged accounts: Toward a socially structured theory of resources and discourses. *Social Forces* 84(1): 89–114

Garduño H & Hinsch M (2005) *IWRM implementation in South Africa: Redressing past inequities and sustaining development with a view to the future.* Washington DC: World Bank Institute

Garduño Velasco H (2001) *Water rights administration: Experience, issues and guidelines.* FAO Legislative Study 70. Rome: Food and Agriculture Organization

Getzler J (2004) *A history of water rights at common law.* Oxford: Oxford University Press

Godden L (2005) Water law reform in Australia and South Africa: Sustainability, efficiency and social justice. *Journal of Environmental Law* 17: 181-205

Government of South Africa (1996) Constitution of the Republic of South Africa No. 108 of 1996, Republic of South Africa

Government of South Africa (1997a) White Paper on Water Policy. Pretoria: Department of Water Affairs and Forestry

Government of South Africa (1997b) National Water Services Act 108. Pretoria: Department of Water Affairs and Forestry

Government of South Africa (1998) National Water Act 36. Pretoria: Department of Water Affairs and Forestry

Graaff J (2006) The seductions of determinism in development theory: Foucault's functionalism. *Third World Quarterly* 27(8): 1387–1400

Grey D & Sadoff C (2006) Water for growth and development. Thematic Documents of the IV World Water Forum, 2006. Mexico City, Comisión Nacional del Agua

Guillet D (1998) Rethinking legal pluralism: Local law and state law in the evolution of water property rights in Northwestern Spain. *Comparative Studies in Society and History* 40(1): 42–70.

Hajer M (1995) *The politics of environmental discourse: Ecological modernization and the policy process.* Oxford: Oxford University Press

Hall R (2004) A political economy of land reform in South Africa. *Review of African Political Economy* 31(100): 213–227

Hall R (2007) Transforming rural South Africa? Taking stock of land reform. In: L Ntsebeza & R Hall. *The land question in South Africa: The challenge of transformation and redistribution.* Cape Town: HSRC Press: 87–106

Hall R (2009) *A fresh start for rural development and agrarian reform?* PLAAS Policy Brief 29. Cape Town: Programme for Land and Agrarian Studies, University of the Western Cape

Hamann R, Khagram S & Rohan S (2008). South Africa's charter approach to post-apartheid economic transformation: Collaborative governance or hardball bargaining? *Journal of Southern African Studies* 34(1): 21–37

Hanekom D (1998) Agricultural policy in South Africa. Discussion Document. Pretoria: Department of Agriculture and Land Affairs

Hendricks F & Ntsebeza L (2000). The paradox of South Africa's land reform policy. SARIPS Annual Colloquium, Harare, Zimbabwe

Hirsch A (2005) *Season of hope: Economic reform under Mandela and Mbeki.* Pietermaritzburg: University of KwaZulu-Natal Press

Hodgson P (2004) *Land and water: The rights interface.* Livelihood Support Programme (LSP) Working Paper. Rome: FAO: 10

Hodgson, P. (2006) Modern water rights: Theory and practice.

Hunt R & Hunt E (1976) Canal irrigation and local social organization. *Current Anthropology* 17(3): 389–411

Jasanoff S (Ed.) (2004) *States of knowledge: The co-production of science and social order.* International Library of Sociology. London: Routledge

Jonker L (2007) Integrated water resources management: The theory-praxis nexus, a South African perspective. *Physics and Chemistry of the Earth* 32: 1257–1263

Jordan P (2004) The African National Congress: From illegality to the corridors of power. *Review of African Political Economy* 31(100): 203–212

Kahinda J-MM, Taigbenu AE & Boroto JR (2007). Domestic rainwater harvesting to improve water supply in rural South Africa. *Physics and Chemistry of the Earth* 37: 1050–1057

Kaplan TJ (1993) Reading policy narratives: Beginnings, middles and ends. In: F Fischer & J Forester. *The argumentative turn in policy and planning.* Durham: Duke University Press: 167–186

Karar E (2010) Water resources regulation: Key issues and principles. Draft discussion document. Water Research Commission Special Report. Pretoria: WRC

King BH (2005). Spaces of change: Tribal authorities in the former KaNgwane homeland, South Africa. *Area* 37(1): 64–72

Laclau E (1996). The death and resurrection of the theory of ideology. *Journal of Political Ideologies* 1(3): 201–220.

Laclau E & Mouffe C (2001) *Hegemony and socialist strategy: Towards a radical democratic politics*. London: Verso

Lahiff E (2003) *The politics of land reform in Southern Africa*. Sustainable Livelihoods in Southern Africa (SLSA) Research Paper No. 19. Brighton: Institute of Development Studies: 19

Lahiff E (2005) From 'willing seller, willing buyer' to a people-driven land reform. PLAAS Policy Brief 17. Cape Town: Programme for Land and Agrarian Studies: 17

Lawrence P, Meigh J & Sullivan C (2002) *The water poverty index: An international comparison*. Keele Economics Research Paper No. 19. Keele University

Levin R & Mkhabela S (1997) The Chieftaincy, Land Allocation and Democracy. In: R Levin & D Weiner. *'No more tears...': Struggles for land in Mpumalanga, South Africa*. Trenton, NJ: Africa World Press: 153–175

Levin R & Solomon I (1997) National Liberation and Village Level Organization and Resistance. In: R Levin & D Weiner. *'No more tears...': Struggles for land in Mpumalanga, South Africa*. Trenton, NJ: Africa World Press: 175-197

Levin R & Weiner D (1997) *'No more tears...': Struggles for Land in Mpumalanga, South Africa*. Trenton, NJ: Africa World Press

Lewis AD (1934) *Water law: Its development in the Union of South Africa*. Cape Town and Johannesburg: Juta & Co. Ltd

Limpitlaw D, Aken M, Lodewijks H & Viljoen J (2005). Post-mining rehabilitation, land use and pollution at collieries in South Africa. Presented at the colloquium on Sustainable Development in the Life of Coal Mining, South African Institute of Mining and Metallurgy, Boksburg (13 July)

Lind JT & Moene KO (2008) Miserly developments. Draft version of unpublished paper. 8 January 2008, available at http://folk.uio.no/jlind/papers/MiserlyDevelopments.pdf

Lorentzen J & Cartwright A (2006) The impact of trade liberalisation on rural livelihoods and the environment: Land governance, asset control, and water access and use in the Incomati River Basin in Mpumalanga: A case study of sugarcane production past, present, and future. Technical Report. Johannesburg: McIntosh Xaba Associates

Maharaj B & Ramutsindela M (2002) Introduction: Post-apartheid political dispensation: New or old geographies? *GeoJournal* 57: 1–2

Malzbender D, Goldin J, Turton A & Earle A (2005) Traditional water governance and South Africa's 'National Water Act': Tension or co-operation? Paper presented at the international workshop on 'African Water Laws: Plural Legislative Frameworks for Rural Water Management in Africa', Johannesburg, South Africa (26–28 January)

Mamdani M (1996) *Citizen and subject: Contemporary Africa and the legacy of late colonialism*. Kampala: Fountain Publishers

Marais, H (2001) *South Africa: Limits to change – the political economy of transition*. Cape Town: University of Cape Town Press

May J (Ed.) (2000) *Poverty and inequality in South Africa: Meeting the challenge*. Cape Town: University of Cape Town Press

Mehta L (2005) *The politics and poetics of water: Naturalising scarcity in western India*. New Delhi: Orient Longman

Mehta L (2010) *The limits to scarcity: Contesting the politics of allocation*. London: Earthscan

Moriarty P, Butterworth J, Van Koppen B & Soussan J (2004) Water, poverty and productive uses of water at the household level. In: P Moriarty, J Butterworth & B van Koppen. *Beyond domestic: case studies on poverty and productive uses of water at the household level*. Delft: IRC, NRI and IWMI

Morris, DR (1998) *The washing of the spears: A history of the rise of the Zulu nation under Shaka and its fall in the Zulu War of 1879*. New York: Da Capo Press

Munzer SR (1990) *A theory of property*. Cambridge: Cambridge University Press

Nattrass N (1994) Politics and economics in ANC economic policy. *African Affairs* 93(372): 343–359

North D (1990) *Institutions, institutional change, and economic performance*. Cambridge: Cambridge University Press

Nozick R (1974) *Anarchy, state and utopia*. Oxford: Blackwell Publishing

Ntsebeza L (2000) Traditional authorities, local government and land rights. In: B Cousins. *At the crossroads: Land and agrarian reform in South Africa into the 21st century*. Cape Town: Programme for Land and Agrarian Studies

Ntsebeza L (2007) Land redistribution in South Africa: The property clause revisited. In: L Ntsebeza & R Hall. *The land question in South Africa: The challenge of transformation and redistribution*. Cape Town: HSRC Press: 107–131

Ntsebeza L & Hall R (Eds) (2007) *The land question in South Africa: The challenge of transformation and redistribution*. Cape Town: HSRC Press

Nxumalo Z (1997) Action for health in Tonga and Shongwe: Initiative for sub-district support. Technical Report No. 2e. Health Systems Trust

Oomen B (2005) *Chiefs! Law, power and culture in the post-apartheid era*. Oxford: James Currey

Pegram G & Bofilatos E (2005) Considerations on the Composition of CMA Governing Boards to achieve representation. Paper presented at the international workshop on 'African Water Laws: Plural Legislative Frameworks for Rural Water Management in Africa', Johannesburg, South Africa (26–28 January)

Perret S (2002) Water policies and smallholding irrigation schemes in South Africa: A history and new institutional challenges. *Water Policy* 4: 283–300

Pickles J & Woods J (1992) South Africa's homelands in the age of reform: The case of QwaQwa. *Annals of the Association of American Geographers* 82(4): 629–652

PLAAS (2004) Land reform in Mpumalanga. Unpublished document made available to the author.

Pollard S, Moriarty PS, Butterworth JA, Batchelor CH & Taylor V (2002) *Water resource management for rural water supply: Implementing the Basic Human Needs Reserve and licensing in the Sand River Catchment, South Africa*. Water Households and IRural Livelihoods (WHIRL), Working Paper No. 6, University of Greenwich. Available on http://www.nri.org/WSS-IWRM/reports.htm

Pollard S & Du Toit D (2005) Achieving Integrated Water Resource Management: The Mismatch in Boundaries between Water Resources Management and Water Supply. Paper presented at the international workshop on African Water Laws: Plural Legislative Frameworks for Rural Water Management in Africa, Johannesburg, South Africa (26–28 January)

Polzer T (2007) Local government and service provision in border areas: Managing migration and facilitating regional integration. Forced Migration Studies Programme Discussion Brief, Johannesburg, University of the Witwatersrand

Ponte S, Roberts S & Van Sittert L (2007) 'Black Economic Empowerment', business and the State in South Africa. *Development and Change* 38(5): 933–955

Pott A, Hallowes J, Mtshali S, Mbokazi S, Van Rooyen M, Clulow M & Everson CA (2005) *The development of a computerized system for auditing real time or historical water use from large reservoirs in order to promote the efficiency of water use.* WRC Report No. 1300/1/05. Pretoria: Water Research Commission

Prinsloo, L (2009) Acid mine drainage in SA's Wits Basin poses significant threat. *Mining Weekly*, from http://www.miningweekly.com/article/western-utilities-2009-09-09

Rangan H & Gilmartin M (2002) Gender, traditional authority, and the politics of rural reform in South Africa. *Development and Change* 33(4): 633–638

Rayner S (2003) Democracy in the age of assessment: Reflections on the roles of expertise and democracy in public-sector decision-making. Social Science Research Council South Asia Regional Fellowship Program Annual Conference, Kathmandu, Nepal, Social Science Research Council (SSRC) (4–12 January)

Rein M & Schön D (1993) Reframing policy discourse. In: F Fischer & J Forester. *The argumentative turn in policy analysis and planning*. Durham: Duke University Press: 145–167

Republic of South Africa (2002) Broad-based socio-economic empowerment charter for the South African mining industry. Pretoria: Department of Minerals and Energy

Roe EM (1991) Development narratives, or making the best of blueprint development. *World Development* 19(4): 287–300

Roe EM (1994) *Narrative policy analysis: Theory and practice*. Durham: Duke University Press

Rose CM (1994) Possession as the origin of property. In: CM Rose. *Property and persuasion: Essays on the history, theory and rhetoric of ownership*. San Francisco: Westview Press.

Rose CM (1994a) Energy and efficiency in the realignment of common law water rights. *Property and persuasion: Essays on the history, theory, and rhetoric of ownership*. San Francisco: Westview Press: 163–196

Rose CM (1994b) *Property and persuasion: Essays on the history, theory, and rhetoric of ownership*. San Francisco: Westview Press

Roth D, Boelens R & Zwarteveen M (Eds) (2005) *Liquid relations: Contested water rights and legal complexity*. New Brunswick, New Jersey: Rutgers University Press

Ruiters G (2002) Race, place, and environmental rights. In: DA McDonald. *Environmental justice in South Africa*. Athens, OH: Ohio University Press: 112–126

Saleth RM & Dinar A (2000) Institutional changes in global water sector: trends, patterns, and implications. *Water Policy* 2: 175–199

Saleth RM & Dinar A (2004) Water institutional reforms: Theory and practice. *Water Policy* 7: 1–19

Salgado I (2009) Acid mine water is a ticking bomb: Researchers warn polluted water from collieries may render Mpumalanga a wasteland. *Business Report* (30 September 2009)

Schreiner, B, Tapela B & Van Koppen B (2010) Water for agrarian reform and rural poverty eradication: where is the leak? Paper presented at the conference Overcoming Structural Poverty and Inequality in South Africa: Towards Inclusive Growth and Development, Boksburg, South Africa (20-22 September)

Schreiner B & Van Koppen B (2002) Catchment management agencies for poverty eadication in South Africa. *Physics and Chemistry of the Earth* 27(11): 969–976

Sen A (2009) *The Idea of Justice.* Harvard: Harvard University Press

Smith A & Stirling A (2006) *Moving inside or outside? Positioning the governance of sociotechnical systems.* SPRU Electronic Working Paper Series 148. Brighton: University of Sussex: 148

Southall R (2004) The ANC and black capitalism in South Africa. *Review of African Political Economy* 100: 313–328

Southall R & Daniel J (2009) *Zunami!: The South African elections of 2009.* Johannesburg: Jacana Media

Sparks A (2003a) *Beyond the miracle: Inside the new South Africa.* Johannesburg: Jonathan Ball Publishers

Sparks A (2003b) *The mind of South Africa.* Johannesburg: Jonathan Ball Publishers

Stats SA (2001) Census report. Available at http://www.statssa.gov.za/census01/html/c2001primtables.asp

Stein R (2000) South Africa's new democratic water legislation: National Government's role as public trustee in dam building and management activities. *Journal of Energy and Natural Resources Law* 18(3): 284–295

Stein R (2006) *Water law in a democratic South Africa: A country case study examining the introduction of a public rights system.* SSRN

Stirling A (2005) Opening up or closing down? Analysis, participation and power in the social appraisal of technology. In: M Leach, I Scoones & B Wynne. *Science and citizens: Globalization and the challenge of engagement.* London: Zed Books: 218–231

Swart, B (2006) Investigating the long-term sustainability of sugar cane farming in small-scale growing projects in the Nkomazi area. Unpublished report. Nelspruit: South African Sugar Association.

Terreblanche S (2002) *A history of inequality in South Africa 1652–2002.* Pietermaritzburg: University of KwaZulu-Natal Press

Thompson H (2006) *Water law: A practical approach to resource management and the provision of services.* Cape Town: Juta

Thompson JB (1984) *Studies in the theory of ideology.* Oxford: Polity Press

Throgmorton JA (1993) Survey research as rhetorical trope: Electric power planning arguments in Chicago. In: F Fischer & J Forester. *The argumentative turn in policy analysis and planning.* Durham: Duke University Press: 117–145

Tisdell JG (2003) Equity and social justice in water doctrines. *Social Justice Research* 16(4): 401–416

Tlou T (2005) Resource-directed measures report. Pretoria: Department of Water Affairs and Forestry

Torfing J (1999) *New theories of discourse: Laclau, Mouffe and Žižek*. Oxford: Blackwell

Toye J (2005) The idea of scarcity in historical perspective. Paper presented at the workshop, Scarcity and the Politics of Allocation, an ESRC Science in Society Programme funded workshop held at the Institute of Development Studies, Brighton, Institute of Development Studies (6-7 June)

Turton A & Henwood R (2002) Hydropolitics in the developing world: A southern African perspective. Pretoria: African Water Research Unit

Turton A, Schultz C, Buckle H, Kgomongoe M, Malungani T & Drakner M (2006) Gold, scorched earth and water: The hydropolitics of Johannesburg. *Water Resources Development* 22(2): 313–335

Turton A R & Meissner R (2002) The hydrosocial contract and its manifestation in society: A South African case study. *Hydropolitics in the developing world: A southern African perspective*. Pretoria: African Water Research Unit: 37–61

Turton A R & Ohlsson L (1999) *Water scarcity and social stability: Towards a deeper understanding of the key concepts needed to manage water scarcity in developing countries*. WCA-Infonet Occasional Papers. Rome: FAO

United Nations Development Programme (2006) Beyond scarcity: Power, poverty and the global water crisis. *Human Development Report 2006*. New York: United Nations Development Programme

Van der Berg S (1998) Consolidating South African democracy: The political arithmetic of budgetary redistribution. *African Affairs* 97: 251–264

Van Der Schyff E (2003) The nationalisation of water rights: Deprivation or expropriation? A South African perspective. Paper presented at the 3rd International Water History Association Conference in Alexandria, Egypt (11-14 December)

Van Koppen B (2006) Formal water rights from a poverty and gender perspective: Deceit, discrimination, and dispossession by design? Paper presented at a conference on Water Reform and Access to Water by the Rural Poor, hosted by Danida, Danish Water Forum and Danish Institute for International Studies, Denmark (16 September)

Van Koppen B, Butterworth J & Juma I (Eds) (2005) *African water laws: Plural legislative frameworks for rural water management in Africa*. Compendium of papers presented at International Workshop Johannesburg, International Water Management Institute (26-28 January)

Van Koppen B, Giordano M & Butterworth JA (Eds) (2007) *Community-based water law and water resource management reform in developing countries*. Wallingford, UK: CABI Publishing

Van Koppen B, Jha N & Merrey D (2002) *Redressing racial inequities through water law in South Africa: Revisiting old contradictions?* Comprehensive Assessment Research Paper 3. Colombo, International Water Management Institute

Villa-Vicencio C & Ngesi S (2003) South Africa: Beyond the 'miracle'. In: E Doxtader & C Villa-Vicencio. *Through fire with water: The roots of division and the potential for reconciliation in Africa*. Trenton, NJ: Africa World Press

Vink N & J Kirsten J (2000) *Deregulation of agricultural marketing in South Africa: Lessons learned*. FMF Monograph No. 25. Pretoria, University of Pretoria

Von Benda-Beckmann F (2001) Legal pluralism and social justice in economic and political development. *IDS Bulletin* 32(1): 46–57

Waalewijn P (2002) Squeezing the cow. MSc thesis. Wageningen: Wageningen University
Wagenaar H & Cook NSD (2003) Understanding policy practices: Action dialectic and deliberation in policy analysis. In: MA Hajer & H Wagenaar. *Deliberative policy analysis: Theories of institutional design*. Cambridge: Cambridge University Press: 139–171
Water Laws Workshop (2005) Plenary statement of participants at the African Water Laws Workshop. Johannesburg (26–28 January)
Watson TJ (1995) Rhetoric, discourse and argument in organizational sense making: A reflexive tale. *Organization Studies* 16: 805–821
Wester P & Warner JF (2002) River basin management reconsidered. In: AR Turton. *Hydropolitics in the developing world: A southern African perspective*. Pretoria: African Water Research Unit: 61–73
Wolfe S & Brooks DB (2003) Water scarcity: An alternative view and its implications for policy and capacity building. *Natural Resources Forum* 27: 99–107
Woodhouse P (1997) Hydrology, soils and irrigation systems. In: R Levin & D Weiner. *'No more tears....': Struggles for land in Mpumalanga, South Africa*. Trenton, NJ: Africa World Press: 75–97.
World Bank (2003) Water Resources Sector Strategy. Washington DC
Zimmerman EW (1951) *World resources and industries*. New York: Harper and Bros

Synne Movik currently holds a postdoctoral position in Global Environmental Governance at the Institute of International Environment and Development Studies, Norwegian University of Life Sciences. Prior to this, she worked on water and sanitation issues with the STEPS Centre at University of Sussex. She received a DPhil from the Institute of Development Studies, also at the University of Sussex, in 2008. The focus of her research was on water policy, water use rights, allocation and governance in South Africa, which provides the basis for the book.

Index

A

Abrams, Len 32
abstraction 63, 81, 107
    determining 124–127
    rights and Afrikaner Nationalism 80
access
    and apartheid 17
    doctrines 1–3
acid mine drainage (AMD) 68–69, 87
adaptive capacity 50, 75n2
administrative rights system 128
affirmative action 16, 26, 52
African National Congress (ANC) 7, 22–26
African Water Issues Research Unit (AWIRU) 55, 56
African Water Laws conference (2005) 59–60
Afrikaner nationalism 7, 16–17, 79–83
    Broederbond 16, 79–80
    hydraulic mission 41, 79, 141, 143
    see also apartheid; Boers
agriculturalist/livelihoods perspective on Water Allocation Reform (WAR) 59–62, 137, 144–145
agriculture 21, 23
    Inkomati area 86
    Land Redistribution for Agricultural Development (LRAD) 94, 103, 132n10
    Strategic Plan for Agriculture 28
    and Water Allocation Reform (WAR) 35, 59–62
    and Water Authorisation and Management System (WARMS) 50–51
    see also commercial agriculture; emerging farmers; small-scale farming; subsistence farming; sugar farming
Agri South Africa (AgriSA) 28, 106
allocation *see* water allocation
AMD (acid mine drainage) 68–69, 87
ANC (African National Congress) 7, 22–26
Anglo-Boer War 13, 16, 79
apartheid 7–8, 14–15, 16–17, 80–83
    see also Afrikaner nationalism
aquifers 30
    contamination of 69
Asmal, Kader 31, 33
Association for Water and Rural Development (AWARD) 38, 56, 93
assurance of supply 39, 84, 116, 116–117
Australia 32–33
availability of water and Water Allocation Reform (WAR) 52
AWARD (Association for Water and Rural Development) 38, 56, 93
AWIRU (African Water Issues Research Unit) 55, 56

B

bakaNgwane people 81–82
Bantu Authorities Act (No. 68 of 1951) 17, 81
Bantustans 17–18, 45n6
    and existing lawful use 41–42
    KaNgwane 81–83, 82, 93–94, 116
Bantu Trust and Land Act (No. 18 of 1936) 79
Bate, R & Tren, R 16, 20, 21
BBBEE (broad-based black economic empowerment) 27–28, 71

BEE (black economic empowerment)
    26–27, 70
Bell, Judge 20
black economic empowerment (BEE)
    26–27, 70
Boers 12–14, 15–16, 58
    Anglo-Boer War 13, 16, 79
    in Inkomati 79–80
    see also Afrikaner nationalism
broad-based black economic empowerment
    (BBBEE) 27–28, 71
Broederbond 16, 80
Bundy, C 15–16
bureaucracy 61, 149

C
cadastral data 123–124
capacity constraints 105–106, 111–112,
    147–148
capital 24, 27, 28–29, 86
capitalism 27, 27–29
    see also neoliberalism
CASP (Comprehensive Agricultural
    Support Programme) 120
Catchment Management Agencies (CMAs)
    8, 33, 36, 148–149
    and Department of Water Affairs and
        Forestry (DWAF) 130
    see also Inkomati Catchment
        Management Agency
Catchment Management Committees 92
catchments
    and allocation 64
    Inkomati 86, 90
    mining in 69
    pollution of 69–70
    stressed 43, 112
    and Water Allocation Reform (WAR) 64

CGIAR (Consultative Group on International
    Agricultural Research) 55
chiefs 12, 13–15
    see also traditional authorities
CLaRA (Communal Land Rights Act (No. 11
    of 2004)) 26, 85, 102–103, 145
CMAs see Catchment Management Agencies
coal mining 69
    see also mining
collateral 101
colonialism 7, 12–16, 20–21, 73–74, 79
    Anglo-Boer War 13, 16, 79
commercial agriculture 28
    and emerging farmers 106–112
    Inkomati Water Management Area
        106–109
    and Water Allocation Reform (WAR) 61,
        70–71
    see also under agriculture
commodification 3
common law 19–20, 21
Communal Land Rights Act (No. 11 of
    2004) (CLaRA) 26, 85, 102–103, 145
compensation
    contested and the National Water Act
        (NWA) 44–45
    and Water Allocation Reform (WAR) 73
Comprehensive Agricultural Support
    Programme (CASP) 120
compulsory licenses 7, 8, 43–44
    authorisation of 130
    Inkomati Water Management Area 119,
        121, 122–124, 131
    and unlawful use 128–129
    and Water Allocation Reform (WAR)
        50–52
    see also licenses
conditional licenses 72

Congress of South African Trade Unions (COSATU) 27
Constitution *see* South African Constitution
Consultative Group on International Agricultural Research (CGIAR) 55
contested compensation and the National Water Act (NWA) 44–45
cooperation and sharing 109–112
    reservoir capacity sharing 76n6
co-production 128
correlative users 128
COSATU (Congress of South African Trade Unions) 27
cotton farmers 115
Council for Scientific and Industrial Research (CSIR) 55, 56, 57, 58, 63
Crocodile River 86, 87
    water yield and demand 88, 89
CSIR (Council for Scientific and Industrial Research) 55, 56, 57, 58, 63
customary system of water rights 2, 3
    and Water Allocation Reform (WAR) 59–60

D

DALA *see* Department of Agriculture and Land Administration
dams 41, 84
    national capacity 30
    storage 105, 107, 108–109, 111
    tailings 69
    see also specific dams
de-agrarianisation 29
decentralisation 8
democratic transformation 22–25
Department for International Development (DFID) 31
    and irrigation 127
    and Water Allocation Reform (WAR) 52, 53–54, 57, 66
    Water and Forestry Support Programme (WFSP) 53
Department of Agriculture (DOA) 91
Department of Agriculture and Land Administration (DALA) 91, 114–115
    allocative authority 116–121
Department of Land Affairs (DLA) 91
Department of Native Affairs (DNA) 81
Department of Water Affairs 47n20
    capacity constraints 147–148
Department of Water Affairs and Forestry (DWAF) 8
    allocative authority 116–121
    and CMAs 130
    and commercial farmers 121–122
    initiating water rights reform 31
    regional offices and municipalities 91, 115
    and SAPWAT 124–125, 128
    and unlawful use 128
    validation and verification of water use 122–130
    and Water Allocation Reform (WAR) 53–54, 56–57
deprivation and expropriation 44–45
deregulation 24, 143
desert by labour 67, 121
development
    economic 23, 72, 142, 147
    rural 147, 150–151
DFID *see* Department for International Development
difaqane 12, 79
discourses
    agriculturalist/livelihoods 59–62
    on allocation 3–7, 62–63, 93, 140
    analysis of 3–6

on efficiency and equity 142–144
on existing lawful users (ELUs) 64–67
on historically disadvantaged individuals
    (HDIs) 64–67, 67–68, 77n16,
    77n17
industrialist/institutionalist 58–59, 61,
    137
and law 62
on scarcity 48–49, 112–114, 140–142
DLA (Department of Land Affairs) 91
DNA (Department of Native Affairs) 81
DOA (Department of Agriculture) 91
domestic use
    and assurance of supply 116, 116, 117
    and General Authorisations 129
    Inkomati Water Management Area
        114–115
    and livelihoods 59, 60
    and the National Water Act (NWA)
        37–39
    and productive use 38
    and reasonable use 38
    and the Reserve 37
    and service delivery 34
Driekoppies dam 84, 87, 93, 117
dryland farmers 115–116
DWAF *see* Department of Water Affairs and
    Forestry (DWAF)

E
ecological Reserve
    feasibility of 38
    and the National Water Act (NWA) 37–38
economic empowerment *see* black economic
    empowerment; broad-based black
    economic empowerment
economic growth 25, 27, 63, 72, 142, 147
economic productivity 8, 70–72

see also productive use
economic risk 40–41, 59, 67–68
economy, political SA 26–30
efficiency 7, 107
    discourse around 141, 142–144
    and equity 70–72, 142–144
    and existing lawful use 128–129
    of irrigation 99, 105, 107, 111–112, 119
Ehlanzeni 102
electricity generation 69, 84, 87, 113–114
ELUs *see* existing lawful users (ELUs)
emerging farmers, Inkomati Water
    Management Area 93–106, 138–139,
    143–144
    and commercial agriculture 106–112,
        120, 143
    and domestic use 115, 117
    irrigation 99, 105, 108
    lack of leverage 111–112
    and macroeconomic policy 143
    and Water Allocation Reform (WAR) 58
    see also emerging farmers; small-scale
        farming; subsistence farming
emerging users 28–29, 58, 67–68, 70–71
employment 67, 68, 72, 121
enforcement and monitoring 64
environment 34–36
    environmental myths 68
    and historically disadvantaged
        individuals 67–68
    and the National Water Act (NWA)
        34–36
    and Water Allocation Reform (WAR)
        67–70
equity 7–8
    and discourse 70–72, 141–142
    and efficiency 142–144

and efficiency in Water Allocation
Reform (WAR) 70–72
water access 28–29, 40
and Water Allocation Reform (WAR)
64–66, 70–72
see also redistribution; redress
equivalential chain 65
Eskom 69, 84, 113
existing lawful use 8, 138
determining 124–130
and the National Water Act (NWA)
40–42
and pollution 67–70
and unlawful use 128–129
validation and verification 122–124
and Water Allocation Reform (WAR)
64–70
Expert Panel 54–56
expropriation 40–41, 73
and deprivation 44–45

F
FDI (foreign direct investment) 58–59
Federation for a Sustainable Environment 69
Figtree 104, 105
flow permits 81, 107
food security 35
forced removals 17, 81, 83
foreign direct investment (FDI) 58–59
forestry 37, 46n20, 53, 87, 114
and Water Allocation Reform (WAR) 53
and water scarcity 114
and water yield and demand 89, 90
fortified rights 138
Fractional Water Allocation and Reservoir
Capacity Sharing (FWARCS) 76n6
Free Basic Services policy 31

FWARCS (Fractional Water Allocation and
Reservoir Capacity Sharing) 76n6

G
GAs *see* General Authorisations
GEAR *see* Growth, Employment and
Redistribution policy
GeaSphere 93
gender relations
Inkomati Water Management Area
103–105
General Authorisations (GAs) 129–130
and the National Water Act (NWA) 39
Gini coefficient 56
GIS data 124
globalisation 27, 61
see also neoliberalism
gold mining 69
see also mining
Government Water Control Areas (GWCAs)
21–22, 36
Komati area 80, 107
Great Trek 12, 79
groundwater 18, 30, 52, 90, 147
Growth, Employment and Redistribution
policy (GEAR) 24
and Water Allocation Reform (WAR)
58, 61
GWCAs *see* Government Water Control
Areas

H
Hajer, M 4
Hall, R 20, 23, 28–29
HDIs *see* historically disadvantaged
individuals
historically disadvantaged individuals
(HDIs)

discourse around 64–67, 67–68, 77n16, 77n17
and efficiency 129
and licensing 52
and Water Allocation Reform (WAR) 52, 64–68, 71–72
homelands *see* Bantustans
human needs Reserve 37
hydraulic mission 41, 58, 79–80, 141, 143

I

identity, social and policy 62–69
illicit trade 22, 120, 127
Incomati River 86
indirect rule 13–15, 16, 17, 81, 85
industrialist/institutionalist perspective on Water Allocation Reform 58–59, 61, 137
industry 21, 43
  see also mining
Inkomati Catchment Management Agency (CMA) 92
  and mine pollution 69
Inkomati Water Management Area 78, 88
  allocative authority 116–121
  and apartheid 79–80, 81–82
  colonisation 78–79
  compulsory licensing, preparations for 122–124
  contested allocations 112
  and discourse 93
  domestic uses 114–115
  dryland farmers 115–116
  emerging farmers 93–106
  established farmers 106–109
  geography 86–90
  historical legacy 79–83

Inkomati Catchment Management Agency (CMA) 69, 92
  irrigation 84–85, 99, 107–108, 111, 116–117, 128
  KaNgwane homeland 81–82
  mining 87
  patronage and paternalism 93
  population 83
  sugar and politics 80–81
  trading of water rights 120–121
  validation and verification of water use 122–130
  water abstractions, determining 124–127
  water availability 86–90
  water management structures 91–93
  water scarcity 112–116
  water sharing 84, 84–85, 109–112
  see also sugar farming, Inkomati
institutionalist discourse 3, 58–59, 61, 137
Integrated Water Resource Management (IWRM) 2–3, 34, 59, 139
  and service delivery 34
  and the state 153–154
International Strategic Perspective (ISP) 89, 90
International Water Management Institute (IWMI) 55, 56, 57, 58, 59
irrigation 30, 84–85, 89
  and assurance of supply 116–117
  determining abstractions 124–127
  efficiency 99, 105, 107, 111–112, 119
  Inkomati Water Management Area 84–85, 99, 107–108, 111, 116–117, 128
  and licenses 43
  Nkomazi Irrigation Expansion Programme (NIEP) 93–94
  and Water Act 1956 35
Irrigation Boards 87, 92

and emerging farmers 111
Inkomati Water Management Area
    107–108, 111, 119, 126–127
and licenses 43
and Water Authorisation and
    Management System (WARMS)
    126–127
Irrigation and Conservation of Water Act
    (no. 8 of 1912) 21
irrigation development
    Inkomati Water Management Area
    84–85
ISP (International Strategic Perspective)
    89, 90
IWMI (International Water Management
    Institute) 55, 56, 57
IWRM *see* Integrated Water Resource
    Management (IWRM)

J
job creation 63, 72
    see also employment
Joint Water Commission (JWC) 84
justice 71, 73–74, 144
JWC (Joint Water Commission) 84

K
KaNgwane homeland 81–83, 82, 93–94, 116
Kasrils, Ronnie 31, 50, 116
Kavin, Hadley 38–39
King, BH 13, 14
knowledge production 55, 57
KOBWA (Komati Basin Water Authority) 84
Komati Basin Water Authority (KOBWA) 84
Komati River 83, 86, 87, 106
    flow permits 107
    pollution 69
    water yield and demand 88, 89, 90

Kruger National Park 34–35, 86, 87
Kwena dam 87

L
labour, migrant 15, 62, 69, 82, 83, 110
land access 12, 15–16, 21, 104–105
    and sugar farming, Inkomati 100–103
Land Act (No. 27 of 1913) 14, 15, 79
Land Redistribution for Agricultural
    Development (LRAD) 94, 103, 132n10
land reform 25, 46n13
    and water reform 60, 73–74, 144–145
    see also land tenure reform
land rights 25, 101–102, 151
    and water rights 19–21, 136
LANDSAT data 124
land tenure reform 25, 101–102
    see also land reform
land transactions and water rights 127
Langeloop 102
language *see* discourse
law and discourse 62
Levin, R 15, 25
licenses
    and bureaucracy 149
    conditional 72
    and the environment 38
    and existing lawful use 41
    and foreign direct investment (FDI)
        58–59
    and general authorisations 39
    Inkomati Water Management Area 109,
        121
    and livelihoods 60–61
    and the National Water Act (NWA)
        42–43, 44–45
    and the Reserve 38
    and use rights 41

INDEX   183

tradable licenses approach 3
see also compulsory licenses
livelihoods discourse 59–62, 137, 144–145
local government 33–34, 91, 114, 115
Lomati River 84, 87, 94, 98, 106–108
Lower Komati River 87
    water yield and demand 88, 89, 90
Lowveld College of Agriculture 125–126
LRAD (Land Redistribution for
    Agricultural Development) 94, 103,
    132n10
Lugedlane Tribal Authority 101–102

M

macroeconomics 26–30
    and the National Water Act 40
    two economies metaphor 29, 40–41
    and Water Allocation Reform (WAR) 143
    see also Growth, Employment and
        Redistribution policy (GEAR);
        Reconstruction and Development
        Programme (RDP); *under*
        economic
MAFF (Ministry for Agriculture, Forestry
    and Fisheries) 151
Maguga dam 84, 87, 90, 94
maize farmers 115
Malthus, Thomas 49
Mamdani, M 12, 13, 14
markets *see* water markets
Masibekela dam 87
Matsamo dam 87
Matsamo Tribal Authority 101, 102, 103
MCCAW (Mpumalanga Co-ordinating
    Committee for Agricultural Water)
    92, 118
mfacane 12, 79
migration 114–115
    labour 15, 62, 69, 82, 83, 110

mining 15–16, 21
    and broad-based black economic
        empowerment (BBBEE) 27
    and pollution 67–68, 87
    and Water Allocation Reform (WAR)
        67–68
    and water demand 89
Mining Charter (2002) 27–28
Ministry for Agriculture, Forestry and
    Fisheries (MAFF) 151
Ministry for Rural Development and Land
    Reform (MRDLR) 151
Mondi 87
monitoring 53, 56, 91, 108, 119, 146
    and enforcement 64
Mozambicans
    hostility towards 83
    migrant 114–115
Mpumalanga Co-ordinating Committee for
    Agricultural Water (MCCAW) 92, 118
Mpumalanga Province 78, 85
Mpumalanga Provincial Government
    Department of Agriculture and Land
    Administration
    see Department of Agriculture and Land
        Administration (DALA)
MRDLR (Ministry for Rural Development
    and Land Reform) 151
Mswati 81, 82
Muller, Mike 33, 38
multinational corporations 27–28
multiple-use systems (MUS) approach 59
municipalities 91, 114, 115
MUS (multiple-use systems) approach 59

N

NAFU (National African Farmers' Union)
    28

National African Farmers' Union (NAFU) 28
National Party 16–17, 28, 80–83
National Water Act (No. 36 of 1998) (NWA) 7, 8, 21, 32, 33
   and Catchment Management Areas (CMAs) 36
   and compensation 44–45
   and compulsory licenses 43–44
   existing lawful use 40–42
   general authorisations 39
   key features and debates 36–45
   and licenses and trading 42–43
   and licensing 43–44
   Reserve and Resource-Directed Measures 37–38
   water use categories 38–39
National Water Policy 32, 49
National Water Resource Strategy (NWRS) 36, 64, 66, 89, 90
   on water yield and demand 88, 90
Natives Land Act *see* Land Act (No. 27 of 1913)
Native Trust and Land Act (No. 18 of 1936) 14
neoliberalism 8, 24, 27, 61
NGOs (non-governmental organisations) 38, 59, 93
   see also specific NGOs
NIEP (Nkomazi Irrigation Expansion Programme) 93–94, 104, 106
Ninham Shands 52, 56, 129
Nkomazi 81, 82
   see also Inkomati Water Management Area
Nkomazi Irrigation Expansion Programme (NIEP) 93–94, 104, 106
Nkomazi/Onderberg Water Action Committee (NOWAC) 121–122

non-governmental organisations (NGOs) 38, 59, 93
   see also specific NGOs
non-scheduled users 125
Nooitgedacht dam 87, 89
normal flow permits 21, 107
NOWAC (Nkomazi/Onderberg Water Action Committee) 121–122
Nozick, R 74–75
Nsikazi 82, 82
Ntsebeza, L 14, 73
NWA *see* National Water Act (No. 36 of 1998)
NWRS *see* National Water Resource Strategy

O
Olifants River 17, 87, 113–114

P
participatory management 148–149
Pass Laws 16–17
permissible water uses 38
Permission to Occupy certificates (PTOs) 14, 101–103, 132n14
permits 33, 43, 107
   see also quotas
PLAAS (Program for Land and Agrarian Studies) 54
policy 33
   and discourse 3–6
   and social identity 62–64, 67
political considerations and water allocation 40–41
political economy SA 26–30
pollution 38, 67–68, 149
poverty
   and general authorisations (GAs) 39, 40
   poverty trap 70–71, 77n19

reduction 7, 29, 103, 151–152
and Water Allocation Reform (WAR) 53, 70–71
water poverty 1, 49
power generation 69, 84, 87, 113–114
pre-colonial era 12
Pricing Strategy 51
prior appropriation doctrine 2, 20–21, 33, 136
private water rights 18–19, 32, 114, 136, 137
privileged accounts 68–70
productive use 8–9, 142–143
and domestic use 38, 60
and ELUs 65, 70, 71
and historically disadvantaged individuals (HDIs) 67, 70, 95, 110
and pollution and waste 69–70
and public interest 71
Program for Land and Agrarian Studies (PLAAS) 54
property rights
and SA Constitution 25, 73, 74
and water rights 1–2, 4–5, 61, 74
see also land rights
proportional use 20, 21
PTOs (Permission to Occupy certificates) 14, 101–103, 132n14
public good 8, 32, 44, 142
public interest 3, 44, 71, 72, 74, 142
public ownership of water 19, 21, 32, 137
public participation 36, 39
pumping 22, 105, 108, 125

Q

quotas 109, 136
commercial farmers 107–108
trading 43, 120, 127
and the Water Act 22
see also permits

R

rainfall tax 37, 87
rainwater harvesting 56
and legislation 38, 39
RDP see Reconstruction and Development Programme
reasonable use 2, 19, 20, 38–39, 137–138
recognised water uses in NWA 37–39
Reconstruction and Development Programme (RDP) 23–24, 26, 31
and Water Allocation Reform (WAR) 61
redistribution 23, 137
and Black Economic Empowerment (BEE) 26–27
and efficiency 140
of land 25
Land Redistribution for Agricultural Development (LRAD) 94, 103, 132n10
and markets 42
and National Water Act 48–49
politics of 150–153
regulation of 148–149
restitution 25, 46n12
and Water Allocation Reform (WAR) 74, 146–147
and Water Law Principles 48
of water rights 3, 7, 8
see also land reform; Water Allocation Reform (WAR)
redress 3, 8, 50, 66, 73
and compulsory licensing 49–50
and foreign direct investment 58
registration of water users 50–52
regulation 148–149
Reserve 7, 31
and contested compensation 44–45
and the National Water Act (NWA) 37–38, 44

reservoir capacity sharing 76n6
reservoirs *see* storage dams
resource capture 31, 58, 73, 141–142
Resource-Directed Measures 37–38
restitution 25, 46n12
rights 8
    of ecosystems 34–35
    see also land rights; riparian rights; use rights; water rights
riparian rights 2, 19–22, 33, 136
    and existing lawful use 40, 42
    Inkomati Water Management Area 107
    and land rights 136
    and land transactions 127
    and licenses 42–43
    and the National Water Act 40
    and the State 137
risk, economic 40–41, 59, 67–68
Roman-Dutch law 18–19, 46n10
run-off 30, 38
rural development 147, 150–151

S

Sabie Sand River 86
    water yield and demand *88, 89,* 90
SACP (South African Communist Party) 27
'safeguard clause' of the NWA 44–45, 73
Sappi 87
SAPWAT model 124–125, 128
SASEX (South Africa Sugar Extension Service) 94, 107, 111
SASRI (South African Sugar Research Institute) 99
satellite data 124
scarcity *see* water scarcity
scheduled users 123, 125
Schedule I of National Water Act (NWA) 38–39, 51
Schreiner, Barbara 32

science and technology and water rights 128
service delivery 29, 30–31, 33–34
    Free Basic Services policy 31
    Inkomati Water Management Area 114–115
    sharing and cooperation 109–112
    reservoir capacity sharing 76n6
SiSwati 82–83
small-scale farming 149
    capacity constraints 106
    and General Authorisations 129
    and irrigation 99, 119
    and multiple-use systems approach 59
    and the National Water Act (NWA) 35
    and the poverty trap 70
    and storage dams 105
    of sugar 93–94, 95, 99
    and Water Allocation Reform (WAR) 70–71
    by women 104–105
    see also subsistence farming
social identity and policy 62–69
social stability 63, 72, 110–111, 141
South Africa
    geophysical characteristics 30
    history 11–17
    political economy 26–30
    water allocation reform 7–9
South Africa Sugar Extension Service (SASEX) 94, 107, 111
South African Communist Party (SACP) 27
South African Constitution (1996) 7–8
    and property rights 25, 73, 74
    and traditional authorities 85
    and water rights 33, 37
South African Sugar Research Institute (SASRI) 99
Sparks, A 12, 13, 14, 16, 79
state, the 139–140

as allocation authority 6–7, 140
and water rights 3, 152–153
storage dams 105, 107, 108–109, 111
storm water 108
Strategic Plan for Agriculture 28
stream flow reduction activity 37, 63
stressed catchments
    Inkomati Water Management Area 112–113
    and licenses 43
stylising characteristics 4–5
subject positioning 5, 64–65, 67, 71, 92
subsistence farming 29
    and Water Allocation Reform (WAR) 71
Sugar Act (No. 28 of 1936) 80
sugar farming 80–81, 93–94
sugar farming, Inkomati
    allocative authority 116–121
    contested allocations 112
    emerging farmers 93–106
    established farmers 106–109
    gender relations 104–105
    land access 100–103
    sharing and cooperation 109–112
    water control and capacity constraints 105–106
    water scarcity 112–116
    yield decline 97, 98, 99
    see also Inkomati Water Management Area
surface water 30, 52, 86
surplus flow 21–22, 107, 108–109
    permits 107
Swaziland 81–82, 84, 86, 113, 116–118
Swazis 83

T

Tenbosch 103
tenure reform see land reform; land tenure reform

Terreblanche, S 12, 13
Thompson, H 18–19, 44–45
TLGFA (Traditional Leadership and Governance Framework Act (No. 41 of 2003)) 26, 85
tradable licenses approach 3
trading of rights 3, 22
    illicit trade 22, 118, 127
    Inkomati Water Management Area 120–121
    and the National Water Act (NWA) 42–43
    temporary and permanent 43
traditional authorities 12, 25–26, 144
    and Bantustans 81
    chiefs 12, 13–15
    and colonialism 13–15
    and the Constitution 85
    and indirect rule 13–15, 16, 17, 81, 85
    and land access 100–105
    and Water Allocation Reform (WAR) 59–60, 61–62
Traditional Leadership and Governance Framework Act (No. 41 of 2003) (TLGFA) 26, 85
transferable water rights see trading of rights
Transvaal Suiker Beperk (TSB) 80–81, 94, 103
tribal authorities see traditional authorities
TSB (Transvaal Suiker Beperk) 80–81, 94, 103
two economies metaphor 29
    and National Water Act 40–41

U

United States of America 20
unlawful use 123, 127, 128–129
unscheduled users 125, 128
Upper Komati Irrigation Board 119

Upper Komati River
  water yield and demand 88, 89, 90
use rights *see* water use rights

V
verification and validation of water use 122–130, 146
Vlakbult 102
Vygeboom dam 87

W
WAMI (Water Allocation Monitoring Index) 145
WAMS (Water Administration and Measurement System) 108
WAR *see* Water Allocation Reform
WARMS (Water Authorisation and Management System) 50–51, 126–127
WARP (Water Allocation Reform Programme) 62–63
Water Act (No. 54 of 1956) 21–22, 35, 80
Water Administration and Measurement System (WAMS) 108
water allocation
  authority 3–7, 140
  authority, departmental 116–121
  and catchments 64
  contested 112
  discourses 3–7
Water Allocation Monitoring Index (WAMI) 145
Water Allocation Reform (WAR)
  agriculturalist/livelihoods perspective 59–62, 137, 144–145
  and compulsory licensing 50–52
  consultation and participation 56–57
  current status 146–147
  and Department for International Development (DFID) 53–54
  and Department of Water Affairs and Forestry (DWAF) 53–54
  drivers of 32–36
  efficiency and equity 70–72, 142–144
  and existing lawful use 64–70
  Expert Panel 54–56
  industrialist/institutionalist perspective 58–59, 61, 137
  initiation of 30–33
  and land reform 60, 73–74, 144–145
  and macroeconomics 143
  policy and social identity 62–64, 67
  and privileged accounts 68–70
  and redistribution 74, 146–147
  registration and compulsory licensing 50–52
  and small-scale farmers 70–71
  and Water Allocation Reform Programme (WARP) 62–63
  see also National Water Act (No. 36 of 1998)
Water Allocation Reform Programme (WARP) 62–63
Water and Forestry Support Programme (WFSP) 131
water resources management (WRM) 53
Water Authorisation and Management System (WARMS) 50–51, 126–127
water, banking of 76n6
Water for Food Movement (WFM) 56
Water for Growth and Development (WfGD) 147
water governance 1–3, 148–149, 152–153
water law drafting team 35, 38, 41
Water Law Principles 32, 35, 48
  and redistribution 48
Water Law Review Panel 32
Water Management Areas (WMAs) 36
  see also Inkomati Water Management Area

water markets 42, 55, 120
    see also trading of rights
water meters 105, 111
water-per-hectare 107
water poverty 1, 49
water quality and quantity 50
Water Research Act 55
Water Research Commission (WRC) 37–38, 42, 55, 76n6
water resources management (WRM) of WFSP 53
water rights
    and capacity constraints 147–148
    and colonialism 18–20
    as *de minimis* rights 39
    determining using science and technology 128
    and discourse 140–144
    evolution of regimes 18–26
    fortified 138
    governance 1–3
    institutionalisation of 3
    and land rights 19–22, 136
    and land transactions 127
    and the National Water Act (NWA) 39
    private 18–19, 32, 114, 136, 137
    redistribution of 3, 7, 8
    Roman-Dutch law 18–19, 46n10
    and SA Constitution 33, 37
    and the state 2–3
    tradable 3
    see also water use rights
water saving incentives 107–108
water scarcity
    contestation around 112–114
    and discourse 49–50, 112–114, 140–142
    first-order and second-order 49–50
    governance 1–3

Inkomati Water Management Area 112–116
    and National Water Act 49–50
water security and the National Water Act (NWA) 39
Water Services Act (No. 108 of 1997) 33
    and rainwater harvesting 39
Water Services Authority 39
water theft 108
water trade *see* trading of rights
Water Treaty, SA/Swaziland 84, 84, 116–118
water tribunal 128
water use
    categories of National Water Act (NWA) 37, 38–39
    determination of 124–127, 146
    see also use rights
Water Use License Application Tracking System (WULATS) 149
water use rights 1–2, 4–5, 8
    evolution of regimes 18–30
    and licenses 42–43
    tradable 3
    see also National Water Act
Water Users Associations 107
weirs 87
WfGD (Water for Growth and Development) 147
WFM (Water for Food Movement) 56
WFSP *see* Water and Forestry Support Programme
WMAs *see* Water Management Areas
women small-scale farmers 104–105
WRC (Water Research Commission) 37–38, 42, 55, 76n6
WRM (water resources management) of WFSP 53
WULATS (Water Use License Application Tracking System) 149

Y

yield
    decline, sugar crops 97–98, *98*
    and demand, water *88, 89, 90*
    models, water 76n6

Z

Zuma, Jacob 150